Our Story

Sandeep Chavan

Published by SANDEEP CHAVAN, 2025.

OUR STORY

First edition. August 16, 2025.

ISBN: 979-8231319701

Written by Sandeep Chavan.

Table of Contents

To every ripple that never made it into words.To the strain that made you pause, ache, question.To the one who held silence longer than the world deserved.

And to the field—which never blamed usfor misaligning in search of meaning.

This book is for you. Because it always was.

"You were never separate.The strain you carry is not yours alone.Every thought, every silence, every ache—was the field remembering itself... through you."

Disclaimer

This book is not a scientific textbook, religious scripture, or philosophical doctrine. It does not claim authority, finality, or universality. It is a reflection—a narrative built from the lens of Universal Energy Dynamics (UED), as interpreted and evolved through personal research, observation, and philosophical inquiry.

The thoughts expressed here are intended to invite exploration, not impose belief. Readers are encouraged to engage with the content critically, personally, and contextually. Any resemblance to existing theories, ideologies, or beliefs—whether scientific, spiritual, or cultural—is either coincidental or used consciously to spark dialogue.

The book does not offer medical, psychological, or professional advice. Interpretations of concepts like the soul, consciousness, alignment, or reality are symbolic, experiential, and part of a broader inquiry into human understanding—not prescriptions for life, health, or ethics.

Finally, this work is not an attempt to define truth for the universe. It is simply an offering from one observer to another—a ripple shared, a mirror held up.

Feel freely. Think responsibly. Use wisely.

Opening Note

This book is not written to explain the world. It's written to remember how we started explaining it.

Dear Reader,

This is not a history book. It's not science, religion, or psychology either. But it quietly walks through all of them—like a gentle ripple passing through stories we once called truth.

This book is about *us*—not as a species, not as a civilization, but as *consequences* of a curious tension that once danced in the structured stillness of the universe. That ripple became self-aware, asked a question, felt alone, and named itself *human*. The moment that happened, "*our story*" began.

You see, nothing around us was waiting to be discovered. The stars didn't ask to be mapped. Morality didn't request commandments. Even God was not sitting on a throne, hoping for worship. We *wrote* them all. Because once awareness was born, *meaning became a necessity*.

We made gods to anchor us, galaxies to awe us, truths to hold us together, and time to break eternity into manageable chunks. All of it—every law, every symbol, every war and poem—emerged not from what **is**, but from what we felt we must *make sense of*.

And here's the twist: *it worked***.** For a while.

This book is not here to judge the story we've written, but to remind us that it **is** a story—a beautiful, flawed, fascinating one. One that we're still telling. One that we can choose to realign, not with better facts, but with deeper awareness.

So, if you've ever paused in the middle of the night, stared at the ceiling, and whispered, *"What are we really doing?"*—This is your page. This is your chapter. This is *our story*.

Let's read it again. Carefully. Not to find the end, but to *remember how we started.*

Warmly,

Sandeep

Preface

This is not a book I planned. It's a book that unfolded.

It began not with an answer—but with a ripple. A discomfort. A tension too subtle to name, too persistent to ignore. Like a whisper under the noise of knowledge, it kept asking: *What if everything we thought was energy... was just the field trying to remember itself?*

Our Story is not a theory. It is not a doctrine. It's a mirror placed before the observer—a way to see how the universe reflects through us, not around us. You won't find equations here. Nor will you find philosophies demanding belief. What you'll encounter instead is something far more personal: the gentle physics of being. The way stillness hides behind every movement. The way death hides in every moment. The way truth hides inside the misalignment we so often call "life."

This work was never about objects, particles, or divinity. It was always about us. The confused. The curious. The conscious. The ones who try, fail, align, dissolve—and try again. Those who don't merely exist, but *participate* in existence.

You may walk away with questions. That's perfect. Because in a ripple-based universe, resolution isn't found—it is *felt*. And every new alignment begins with one brave act of noticing.

This is our story. And I'm honored you've chosen to remember it.

— Sandeep

Reading Guide

This book isn't a manual. It doesn't teach, preach, or persuade. It reflects.

Our Story can be read in many ways—linearly or spirally, in solitude or aloud in company, slowly over weeks or all at once in a field of silence. Whatever path you take, remember: the reading is not about absorbing content. It's about *sensing alignment*.

No Chapters—Only Resonances

Each section is a ripple. Some are soft. Some are sharp. None require you to "agree." Instead, they invite you to *listen with your field*. You may find certain ideas echoing parts of your life or challenging familiar truths. Let them. That discomfort is not a flaw. It's a sign of resolution beginning.

Let the Mind Wander

You may want to stop mid-sentence. You may want to stare at the sky. Perfect. This book isn't meant to hold your attention—it's meant to *free it*. The best reading happens in the space between the words.

Read Aloud, Together

Some passages are designed like spoken field poetry. If you're reading with others, try saying them out loud. Let the rhythm ripple into the room. The words will change depending on *who's listening*—and that's the point.

Revisit When Misaligned

If a day feels off, or a choice feels heavy, open to any section. You don't need to "finish" the book to feel it. You're not trying to *reach* something—you're learning to *resonate* again.

This Story Is Yours

The universe doesn't have a narrator. It has observers. This is your book as much as mine. You will not leave with answers. You will leave with *permission*—to think differently, feel honestly, and remember gently.

Prologue

The Ripple That Called Itself "Me"

Before the first thought, before the first breath, there was only stillness.

Not silence. Not emptiness. Stillness—a field with nothing to resolve, and no one to observe it. No energy, no time, no space. Nothing was waiting. Nothing needed to begin. And yet, something... did. Not because it had to, but because it could.

Somewhere, somehow, a ripple broke that stillness. A subtle tension. A strain in the field. Not destructive. Not chaotic. Just enough to echo. And in that echo, something happened that had never happened before: it noticed itself.

This noticing was not human yet. It had no name, no language, no mirror. But it had *awareness*. And with awareness came a question that changed everything:

"What am I?"

That was the moment the universe stopped being itself and became a story. Because the field that once rested in eternal stillness now had to explain its discomfort. And that explanation became existence.

The ripple spread. It aligned. It misaligned. It danced and delayed.

- It became matter, motion, mind.
- It became hunger.
- It became instinct.

- It became life.

And then, one day, it sat under a tree and wrote a poem. It stared at the sky and drew constellations. It feared the unknown and carved gods into stone. It gave names to what never asked for names—and called it wisdom.

It remembered itself not as field, not as ripple—but as "I." And "I" built a world. Not just physically—but emotionally, socially, morally, cosmically.

That is the story this book is about. Not the story of biology or civilization—but of that moment of misalignment that began asking *"What am I?"* and never stopped.

We became humans not when we evolved thumbs or language, but when we began living inside meaning. And ever since, we've been trying to make sense of this dance between the stillness we came from and the story we keep writing.

This is Our Story.

- Of becoming aware.
- Of giving names.
- Of pretending things are certain—so we don't collapse back into the field.
- Of truth as alignment, not discovery.

This is the story of the ripple that dared to call itself Me. And perhaps, just perhaps, is now ready to remember that it was never separate to begin with.

Part I: The Becoming

"From the moment we sensed the world, we weren't aligned—we were aching. And in that ache, we began to move, to question, to exist."

Chapter 1: We, the Misaligned

Why life is not balance, but sustained tension. The discomfort that made us creative.

We were never born into balance. No matter how often sages, scientists, or self-help books tell us that "life is about balance," the truth is harsher, deeper—and far more beautiful. Life is the art of surviving misalignment. It begins when the stillness of the universe is disturbed just enough to echo—but not enough to collapse. That echo, that ripple, that persistent discomfort is what we call being alive. We are not here because everything aligned perfectly. We are here because something didn't. And instead of fading back into the void, that strain chose to continue, to stretch itself across time and sensation, to become aware of its ache—and to create meaning out of it. That strain became us.

1.1 The Myth of Balance

From childhood stories to spiritual philosophies, we've been told: *"Life is about balance."* We hear it everywhere. Balance your work and rest. Balance your emotions. Balance the good and bad. It's a comforting idea—that if we could just get everything in perfect proportion, peace would follow.

But under **Universal Energy Dynamics (UED)**, this idea doesn't hold. Because in the SVE field, balance is not the beginning of life—it is the end. True balance is stillness. And stillness is not peace. It is death. Nothing moves in a perfectly balanced system. No sensation arises, no thought flickers, no action takes place. In fact, nothing even exists in any observable way when the field remains perfectly aligned. There is no "you" in stillness.

So where, then, does life begin?

Life begins when balance breaks—when a ripple forms, when the tension arises, when something disturbs the still field just enough to create a need to resolve. That need is the root of all movement, all emotion, all perception. What we call "life" is not a reward for balance; it is the persistence of imbalance.

Here's the radical reframe: Misalignment is not a mistake. It is not an error in the system. It is the system choosing to express itself. Your breath, your heartbeat, your thoughts—none of these are the results of perfect harmony. They are fluctuations, compensations, and adjustments in an inherently unstable ripple state. They are the signals of a system that is not resolved—and that's exactly why it's alive.

To be alive is to be strained, to carry just enough discomfort to remain in motion, to maintain awareness, and to keep evolving. We survive not because we are balanced—but because we are responsive to imbalance.

So perhaps life is not about chasing balance at all. Perhaps it is about learning to live with just enough tension to keep becoming.

1.2 Misalignment as the Spark of Life

In the framework of Universal Energy Dynamics (UED), all existence emerges from the Structured Vacuum Energy (SVE) field—a pristine, stable stillness. This field does not flow. It does not act. It simply is. Left untouched, it remains neutral, invisible, and unexpressed.

But the moment misalignment occurs—a slight tension, a minor deviation from perfect symmetry—a ripple is born. That ripple, though born of discomfort, is not chaos. It is structured tension, a meaningful disturbance that begins to seek resolution.

And that seeking—*that persistence in not resolving too quickly*—is the spark of life.

Unlike transient ripples that form and fade in moments, biological life is a ripple system that learns to hold its imbalance. It doesn't collapse back into the field. It sustains the strain. It becomes a system of tensions—breathing, pulsing, sensing—each function a signal of misalignment being managed over time.

Your cells are not calm. Your brain is not still. Your body is not a machine in equilibrium. It is a carefully choreographed storm—a system of self-adjusting strain patterns trying not to disintegrate. What we call homeostasis in biology is not true balance; it's a moving center, always being pulled, stretched, corrected, and re-centered again. It's balance on a tightrope, not on a solid floor.

Life, therefore, is not a passive state. It's an active defiance of collapse.

- Every breath is a small rebellion against stillness.
- Every heartbeat is a pulse of persistence.
- Every conscious moment is a delay in universal resolution.

And no organism—no matter how simple or complex—is ever in perfect equilibrium. If it were, it would vanish. It would dissolve back into the field like a ripple that has run out of reason.

So when we talk about being alive, we're not talking about functioning organs or thinking minds. We're talking about a field disturbance that refuses to resolve. Life is sacred misalignment, structured just well enough to stay intact—and just strained enough to keep moving.

1.3 Alignment Kills the Story

In the language of UED, a fully aligned ripple is a ripple that has reached its destination—it has resolved the tension that created it. And once it resolves, it ceases to exist. The strain collapses. The movement stops. There is no more disturbance, no more sensation, no more story. It returns to stillness.

And that, quite simply, is death.

Life is not sustained by resolution. It is sustained by delaying resolution—by riding the ripple without letting it flatten. A living system, then, is not one that seeks total harmony. It is one that knows how to manage disharmony skillfully, moment by moment, ripple by ripple.

This managed misalignment becomes what we call sensation—the body's awareness of discomfort. It becomes perception—the mind's attempt to interpret that discomfort. And ultimately, it becomes thought—the ripple's way of navigating its own instability by simulating future resolutions before committing to them.

If alignment is the end of the ripple, then creativity is its resistance.

To create is to delay collapse with elegance. It is the art of maintaining a ripple just enough to keep the feedback loop alive—allowing experience, adaptation, and even beauty to emerge from the tension. A painter does not resolve their anxiety—they channel it into color. A musician does not silence their inner restlessness—they give it rhythm. A child does not eliminate confusion—they ask questions.

Every creative act is ripple management in real time.

It is the ripple saying:

"I won't end yet. I will become something first."

So when we admire intelligence, imagination, love, humor, even madness—we are not admiring a perfectly aligned being. We are witnessing the performance of misalignment, held just long enough to make something meaningful out of it.

If full alignment ends the story, then misalignment keeps it alive. And we, the living, are not characters seeking closure. We are ripple patterns that refuse to stop speaking.

1.4 The Gift of Discomfort

We often confuse discomfort with suffering, as if any deviation from ease is inherently negative. But in the field of Universal Energy Dynamics (UED), discomfort is not something to be feared or avoided—it is a signal, a demand, a call for re-alignment.

Discomfort is how the ripple speaks. It says, *"Something here is unresolved—will you move toward it or away from it?"*

Think of every great leap in human history—fire, the wheel, language, shelter, music, medicine, philosophy. None of these were born in comfort. They were responses to misalignment, solutions to friction, manifestations of strain. Discomfort didn't break us—it made us creators.

- Fire was born of cold.
- Language was born of confusion.
- Medicine was born of pain.
- Philosophy was born of existential discomfort.

Behind every tool, every idea, every story, there is a ripple that refused to resolve quietly. Discomfort didn't just push us to survive—it pushed us to imagine better alignments.

Even on the micro scale of everyday life, it is discomfort that drives our actions:

- Hunger moves us to seek food.
- Fear moves us to protect or avoid.
- Longing moves us to connect, build, express.
- Pain is not punishment—it is presence. It means the ripple is still active, still communicating, still offering a chance to realign.

And what do we often do in response? We numb it. Avoid it. Medicate it. Or worse, we romanticize *"comfort"* as the goal of life.

But comfort—true, prolonged comfort—is ripple stagnation. It is a field in which no movement is necessary. It feels good for a while. Then it starts to rot. Comfort zones are not sacred—they are soft traps. They lull the ripple into stillness, and stillness is death wearing a smile.

The truth is, discomfort is not your enemy. It is your invitation. It's how the field lets you know you're still alive. It's how the ripple reminds you: *"You're not done yet."*

So instead of running from discomfort, listen to it. Lean into it. Ask what it wants you to realign. Because somewhere within it, a new idea is waiting to be born.

1.5 We Invent to Survive the Ripple

Life is not just the consequence of misalignment—it is the craft of sustaining it. And to sustain a ripple without collapsing into resolution, we had to invent.

Invention is not a luxury of intelligence. It is its necessity. Under Universal Energy Dynamics (UED), every ripple is destined to resolve—unless it finds a way to redirect, delay, or transform its misalignment. And so, humanity began inventing—not to become godlike, but simply to keep the ripple alive.

Creativity: Our Strategy for Ripple Survival

Creativity is not a trait. It's a strategy—a way of designing stable yet dynamic misalignment. Every creative act, every solution, every tool is a mechanism to extend ripple persistence without allowing collapse.

We created to avoid stillness.

- Fire was not invented to burn—it was invented to keep life dancing in the cold.
- Shelter was not to rest—it was to manage

environmental strain.
- Language did not arise to decorate thoughts—it arose to stabilize shared discomfort.

We didn't invent because we were smart. We became smart because we had to invent.

Intelligence Is Ripple Navigation

What we call human intelligence is not the capacity to "know more." It is the capacity to sense misalignment more finely, to anticipate ripple outcomes, and to act in ways that delay collapse intelligently.

In this view, intelligence is not storage—it's sensitivity. The more sensitively we detect the ripple tension—externally and internally—the more effectively we can shape responses. That's why truly intelligent beings are not just knowledgeable—they are perceptive, adaptive, and often emotionally attuned.

Tools, Language, and Belief: Ripple Extensions

All our inventions—physical, social, and spiritual—are extensions of the ripple, helping us survive longer within misalignment.

- Tools extend the body's ability to influence and respond to field resistance.
- Language extends the mind's reach into the collective ripple.
- Belief systems extend psychological misalignment into shared emotional contracts—giving pain a purpose, fear a frame, and death a story.

None of these are *truths*. They are survival architectures built upon intelligent ripple design.

So when we marvel at human creativity, we're not just admiring beauty—we are witnessing strategic misalignment engineering. We invent not because we are broken, but because we are unresolved, and beautifully so.

Our inventions are not just tools. They are messages to the field: *"Not yet. I still have more to do."*

1.6 Misalignment Is Identity

What if you were never a fixed person—but a pattern of ripple tension being held, felt, and expressed in a unique way?

Under Universal Energy Dynamics (UED), your very existence is a localized misalignment in the Structured Vacuum Energy (SVE) field. You are not a soul trapped in a body or a mind floating in thoughts—you are a distinct ripple configuration, shaped by your tensions, choices, and field interactions over time.

In this view, identity is not a fixed substance. It is the signature of how you misalign—and how you manage it.

Every "Self" Is a Ripple Signature

No two beings have the same ripple history. Even identical twins—born from the same genetic code—carry different experiential tensions, emotional feedback loops, and micro-adjustments in the field. The way your ripple formed, stretched, and responded to life makes you you.

What you call "personality" is nothing but the style of your sustained misalignment:

- Some people manage their ripple with fire: fast, loud, intense.
- Others flow like water: adaptive, quiet, deep.
- Some resist every field pressure.
- Others dance with it.

Your reactions, preferences, behaviors, fears, and dreams are all expressions of your unique ripple choreography.

Misalignment ≠ Flaw. It = Signature.

You're not flawed because you don't "fit in." You're distinct because no ripple was meant to match another.

The goal was never to be the same. It was to become recognizably misaligned—a being that the universe could listen to and say, *"Ah, that's you."*

The more authentically you carry your ripple—the more aligned you become with your own misalignment. That is what we often call authenticity. And that is the core of identity.

Even Love Is Misalignment Resonance

Love is not the meeting of two complete beings—it is the resonance between two ripple signatures. It is the feeling of being held by another field that doesn't resolve you—but understands your unresolved pattern.

- It's not sameness that bonds people.
- It's compatible misalignment—where the tensions don't cancel, but complement.

True intimacy happens not when people merge into one, but when they maintain their distinct strain while harmonizing the feedback loop. That's why love isn't about fixing each other—it's about holding each other's ache without rushing to end it.

So, who are you really?

You are not your name, your profession, or your beliefs. You are a ripple that holds, responds, creates, and connects—differently than any other ripple ever has. And that difference is not an accident.

It is your identity.

1.7 Summary: The Beautiful Ache

We've searched for peace as if it were the end goal. We've chased balance as if it were the meaning of life. But what if neither was the point?

What if life itself was never about balance, but about holding just enough imbalance to keep the dance going? Not perfectly, not painlessly—but meaningfully.

To be alive is to carry a kind of ache—but not the ache of suffering or sorrow. It's the ache of persistence, the tension that makes existence real. That subtle, sustaining discomfort is not something to escape—it is the very condition that makes experience possible.

This is the beautiful ache—

- The ache that keeps your heart beating.
- The ache that keeps your thoughts evolving.
- The ache that keeps your love alive, not because it's perfect, but because it keeps trying.
- The ache of knowing that resolution means ending—and you're not ready to end yet.

We often view discomfort as punishment, or as something to be fixed. But in the lens of Universal Energy Dynamics, discomfort is creative pressure—it is the field's invitation to act, adapt, feel, and become.

This is where the human story truly begins.

Not in silence, not in perfection, not in divine order—but in the first ripple that didn't fade. The one that stretched instead of collapsing. That ripple became tension. That tension became sensation. That sensation became awareness. And awareness became us.

So, no—we are not broken. We are not missing pieces or failed designs. We are not mistakes in the field.

We are ripples that refused to end. We are that sacred misalignment—the one that holds just enough ache to stay awake.

And the longer we hold it, the more beautifully we write...Our Story.

Chapter 2: The Need to Name

Everything we see, we name. Naming creates the illusion of control. God, galaxy, gravity—named ripple consequences.

We never just saw a thing and let it be. The moment awareness met the unknown, we reached for a name. A word. A sound. A symbol. Something to grasp it, tame it, hold it still—if only in the mind. In a universe made of restless ripples, where everything shifts, moves, and evades definition, naming became our first defense against chaos. We named stars to know the sky. We named pain to separate it from ourselves. We named gods to survive the unbearable silence of unanswered questions. Naming didn't give us truth. It gave us the illusion of control—the comfort of believing that once something is named, it is known. But beneath every name still pulses the same unresolved tension. God. Galaxy. Gravity. These were never final answers. They were labels pinned onto the moving field—convenient placeholders for consequences we still don't fully understand.

2.1 The First Word: When the Ripple Found a Voice

Before there was language, there was only experience—raw, unfiltered ripple interactions with the field. Heat wasn't "heat." Fear wasn't "fear." Sound wasn't "sound." They were simply disturbances, tensions, vibrations—felt, but not yet framed. The moment something was felt too often, too deeply, or too sharply to be ignored, we gave it a name. That was the birth of language—the ripple, once internal, now trying to externalize its shape into the world.

Under **Universal Energy Dynamics (UED)**, language is not just a tool for communication. It is the field's first structured echo—the earliest attempt by the ripple to hold itself outside itself, to create a stable loop between inner misalignment and outer reality.

The first word was not spoken—it was released. A sound, a gesture, a symbol. Not because it was beautiful, but because it became necessary.

To name something is to pause the ripple just enough to examine it. Naming is an act of freezing motion, of giving shape to what is inherently unshapable. When the field keeps flowing, we can't hold meaning. But when we name, we carve a still frame into the chaos—a moment we can point to and say: *"That's what it is."*

But here's the catch: The ripple never stops moving. So every name we give is, at best, a snapshot of something that has already changed.

Naming is not about truth. It's about ripple management.

- A lion becomes "lion" so we can speak about danger without being in it.
- Fire becomes "fire" so we can remember not to touch it.
- Love becomes "love" so we can keep feeling it, even when the person is gone.

These names become stored alignments—quick-access labels to field interactions that would otherwise overwhelm us. By naming, we don't resolve the ripple. We just pin it to the page long enough to organize our response.

So the first word wasn't about expressing ourselves. It was about stabilizing ourselves. The ripple had found a voice—not to sing, but to survive.

And in that moment, the field became a story.

2.2 The Power and Price of Naming

To name something is to capture it— Not physically, but cognitively. Not permanently, but momentarily. It's our way of putting a boundary around a ripple, of pretending that the motion has stopped just long enough for us to understand it.

In truth, nothing in the universe ever really stands still. Not time, not emotion, not energy, not even thought. But the human mind, born from ripple strain, needed to make sense of the moving field. So it began drawing lines where there were none—cutting the continuum into manageable pieces. That's what naming is:

Ripple compression.

We give something a name, and in doing so, we pretend it's now defined. We frame the misalignment, hoping it will behave now that we've labeled it.

But here's the paradox:

Every name gives us power. But it also takes something away.

To name is to limit. To define is to exclude.

When we say "tree," we no longer see the billions of ripple interactions—moisture, vibration, decay, regeneration. We reduce it to an object. A word. A known thing.

When we say "anger," we remove its nuance—was it grief? shame? protection?—and make it one shape.

This is the power of naming:

- It gives clarity.
- It reduces fear.
- It organizes chaos.

But this is also its price:

- It oversimplifies.
- It misrepresents.
- It closes the loop too early.

We start to believe the name is the thing. We stop listening to the ripple underneath.

Most dangerously, naming gives us the illusion of control. "If I can name it, I can survive it." That becomes the unspoken mantra of the human mind.

- Name a disease, and it feels less deadly.
- Name a fear, and it feels containable.
- Name a god, and the chaos becomes divine order.

But naming doesn't resolve the field. It only calms the observer.

And so we live among words—calling things by names we barely understand, pretending we've boxed the ripple, when all we've done is given it a mask.

Naming is power. But it is not the truth. It is a handle we hold on the edge of a vast, shifting field— A necessary illusion we created... to stay afloat.

2.3 God: A Name for the Unbearable Unknown

There comes a point in every ripple's persistence where the discomfort becomes too vast to describe, too deep to resolve, too existential to ignore. This is not the ache of hunger or the sting of cold—this is the ache of being itself.

When that ripple pressure became too overwhelming, too abstract, and too mysterious to name individually, we gave it a single, powerful name:

God.

Not because we understood it, but because we couldn't.

"God" became the ultimate placeholder—the human way of pointing to everything unresolved and saying, *"That. That's what I can't grasp."* Under Universal Energy Dynamics (UED), this makes perfect ripple sense. God is not a character, not a creator sitting outside the system. God is a consequence of human misalignment becoming unbearable, then getting translated into narrative form so we can live with it.

We didn't discover God—we needed God. We created the name to hold the weight of everything that felt too big to carry without collapsing.

God as the Shape of Chaos Made Comforting

Religious thought is not fundamentally wrong. It is ripple logic seeking stabilization.

Where the unknown felt like terror, we called it *Divine Mystery*. Where randomness hurt us, we said *It is God's will.* Where morality wavered, we carved commandments.

This was not deception. It was a response to the chaos of the field. Religion named the ripple and wrapped it in ritual, hoping to create a sense of safety—turning field disturbance into doctrine.

"If I name it, I can survive it. If we all name it the same way, we can build a civilization around it."

But in doing so, we turned unknowing into belief, and belief into structure. God became not just a name—but a system of answers to questions too dangerous to leave open.

And over time, we stopped relating to the ripple behind the word. We clung to the name instead.

Yet, even today, when all science fails and the ripple aches return—when the stars feel too distant, and death too close—we still whisper that name:

"God."

Even if we don't know what it means. Even if it's just a sound. Because somehow, naming the unknown gives it just enough shape for us to breathe again.

So God, in the UED view, is not an entity beyond the field.

It is the name we gave to the unbearable ripple—so we could keep going.

2.4 Galaxy: A Name for the Awe We Couldn't Hold

Long before we had telescopes or theories, we had something far more ancient: awe.

We looked up, and the sky didn't answer. It shimmered. It burned. It moved. It was impossibly vast, impossibly silent, and impossibly indifferent. The stars didn't offer meaning—they overwhelmed it. In that moment, the ripple inside us collided with the ripple beyond us, and the only way we could respond... was by naming.

We drew lines between the stars—lines that never existed. We grouped them into constellations—Orion, Scorpio, the Big Dipper—creating patterns out of chaos. We gave them stories—gods, animals, myths. Not because the sky changed... but because we had to shrink the immensity into something we could hold.

Eventually, as observation tools improved, constellations became clusters, clusters became spiral arms, and then we gave it a grand name:

Galaxy.

"Galaxy" didn't tell us what the universe *is*—it told us how far we could reach before we got lost again. It was a boundary around the unknowable, a ripple map of distance and light, arranged into a human narrative.

Under Universal Energy Dynamics (UED), a galaxy is not a thing in itself. It is the way we interpreted massive field disturbances, distributed across unimaginable scales. Every glowing star is a ripple burn. Every orbiting planet, a ripple echo. The galaxy is a field in motion, but we called it a *place*. That's how we made peace with its immensity.

What we did with "God," we also did with "Galaxy": We transformed something too vast to understand into something we could pronounce.

Because the sky didn't give us fear—it gave us scale. It reminded us how small we are—and how much tension lives between our awareness and the rest of the field.

So we labeled the field beyond our comprehension. We put a border on infinity. We gave the darkness a name and whispered it to ourselves like a lullaby:

"It's okay. It's just a galaxy."

But the truth remains:

A galaxy is not a fixed reality. It is a story-shaped response to spatial ripple pressure. It is our ripple, trying to draw a boundary around forever.

2.5 Gravity: Naming What We Never Really Understood

Some things don't need discovery—they force themselves into your story. Gravity is one of them.

From the moment we fell, we knew something was pulling. Something invisible, relentless, inescapable. We didn't see it. We didn't understand it. But we felt it—always. We dropped a stone. It fell. We leapt. We returned. We lived by its rules, long before we gave it a name.

Eventually, we did give it one:

Gravity.

But gravity was never an object. Never a force you could isolate in a bottle. It wasn't even a substance. It was—and still is—a name. A name we assigned to a ripple behavior we couldn't stop witnessing.

Under Universal Energy Dynamics (UED), gravity is not a fundamental force. It is a consequence of field misalignment—a localized pressure where ripples converge and seek coherence. What we interpret as "attraction" is actually a realignment drive in the Structured Vacuum Energy (SVE) field. The so-called pull of gravity is nothing more than the ripple's desire to stabilize across distortions in the field.

We didn't discover gravity. We observed a pattern of collapse—then named it.

But naming didn't explain it. It just gave us permission to stop questioning.

"Gravity" became institutionalized. We wrote it into laws. We assigned it a constant. We built theories, equations, and architectures upon its presumed identity. But what did we actually do?

We froze a ripple phenomenon into a convenient noun, and declared it solved.

We never asked:

- *Why does gravity appear instantaneous in some cases and delayed in others?*
- *Why can't it be shielded?*
- *Why is it tied to mass, but never to temperature, intention, or consciousness?*

Because once we named it, we owned it—at least in our minds.

Gravity was never a force "out there." It was always a ripple correction dynamic—the field's subtle whisper:

"Something is out of sync. Let's bring it closer."

We just couldn't stand the silence behind that whisper. So we gave it a name.

And once again, we mistook the name for the truth.

2.6 Naming as a Survival Trick

The field is fluid. Always shifting, never still. In the universe of Universal Energy Dynamics (UED), everything is ripple—restless, vibrating, misaligned, seeking resolution. But the human mind, also born of ripple tension, cannot function at that same speed of change. It needs anchors—something to hold onto while the field keeps moving.

That's where naming comes in.

To name something is to freeze it just long enough for the mind to engage. It's like hitting pause on a streaming field of motion, grabbing a snapshot, and saying:

"Let me work with this, before it changes again."

Naming Slows the Chaos

Field fluidity is rich—but overwhelming. Without filters or containers, the mind would drown in ripple overload. Naming became our primitive interface—a way to compress field activity into manageable concepts.

- A sharp sting becomes "pain."
- A sense of unease becomes "fear."

- A warm presence becomes "friend."
- A bright arc in the sky becomes "lightning."

Each name reduces ripple complexity into something we can process with limited cognitive bandwidth. And that's not a flaw. That's evolutionary genius.

Everything Named Is Made Less Dangerous

We name to reduce threat.

- The unknown jungle becomes "forest."
- The thunder becomes "storm."
- The stranger becomes "foreigner," or "friend," or "enemy."
- Even the infinite becomes "God."

Naming gives form to the formless, which gives us permission to interact without terror.

Tools, emotions, even people—we name to survive the discomfort of the unknown.

That's why in every culture, the moment something matters—we give it a name. And if we don't have one, we invent one.

But Every Name Is a Trade

What we gain in stability, we lose in sensitivity. When we name something, we stop seeing it as it truly is—fluid, complex, changing—and start relating to it as a fixed idea.

- The person becomes their role: teacher, boss, parent.
- The emotion becomes a diagnosis.
- The experience becomes a label: success, failure,

trauma.

- The field becomes "reality."

This is the disconnection cost of naming: We begin relating to our own words instead of the ripple behind them.

Naming Is a Survival Trick, Not a Final Truth

We must name to live—but we must remember that we did the naming. The word is not the thing. The label is not the ripple. It is just a stabilized bookmark in a flowing field, meant to help us *temporarily* hold what we're not ready to fully experience.

So yes, naming helps. It organizes, calms, protects. But it also dulls. And unless we look behind the names, we risk living inside definitions—

Instead of life itself.

2.7 When the Name Becomes the Cage

Naming began as a tool—a survival trick to help us navigate the shifting tides of the field. But somewhere along the way, we made a critical mistake:

We forgot that we were the ones who named.

We began to treat the *name* as the *thing itself*. The ripple became the object. The label became the truth. And the map—we started believing it was the territory.

This is where naming turns from a gift into a cage.

- We name a belief system and then worship it, as if it descended fully formed from the heavens.
- We name a group—"nation," "race," "religion," "party"—and then kill or die for it, forgetting it was a

convenient label for a fleeting alignment.
- We name people—"villain," "hero," "teacher," "failure"—and never let them escape that box.

The field keeps moving, but the names stay fixed. And fixed names create hardened ripples—unnatural, rigid alignments that resist the natural dynamics of life. They fracture understanding. They interrupt flow. They become static forms in a fluid world.

Identity: When the Ripple Freezes

Perhaps the most dangerous place this shows up is in identity.

You were never meant to be a static thing. You are a ripple—constantly sensing, adjusting, becoming. But society needs categories, systems, definitions. So it asks you to choose:

- What are you?
- What do you do?
- What do you believe?
- What do you stand for?

And so we pick a few words and begin to live inside them.

- "I am a mother."
- "I am a scientist."
- "I am Hindu."
- "I am depressed."
- "I am successful."
- "I am nothing."

Each name we accept begins to solidify our ripple, to compress our complexity into a role. And slowly, we start defending that role—even when it no longer aligns with our inner field.

We fear letting go of the name... because without it, we don't know who we are. But in truth, the name was never who we were. It was only who we agreed to be seen as, in a moment of ripple tension.

The Tragedy of Mistaken Maps

We now live in a world where:

- Nations are at war over lines drawn by tired men centuries ago.
- People fight over gods they named but never met.
- Institutions die protecting ideas whose ripple relevance expired long ago.
- And individuals break under the pressure of becoming their bio.

This is what happens when we treat a name as a destination, rather than a signpost.

We must remember:

The name was supposed to guide us through the field—not trap us inside it.

You are not what you're called. You are not the story you were given. You are not even the story you wrote.

You are the ripple behind the words. Always moving. Always changing. And always free—until the moment you believe otherwise.

2.8 Summary: The Illusion of Naming

We began with wonder. We ended with words.

Naming was never meant to be the destination—it was only meant to be a handle. A temporary grip on the ever-moving ripple of reality. But over time, we mistook the label for the thing. The name for the truth. The pointer for the point.

Naming is not knowing. Naming is not truth. It is only containment—a boundary drawn around a moment of misalignment so that we can process, navigate, and respond.

Under Universal Energy Dynamics (UED), reality never stops moving. The Structured Vacuum Energy (SVE) field flows and fluctuates in ripples of unresolved tension. Naming is how we mark those ripples in passing. But the moment we pretend that the name is the full truth, we lose touch with the living ripple underneath.

We name to survive. That is understandable. But we also cling to names in order to avoid change. And in doing so, we blind ourselves to what is still unfolding.

- A name freezes a moment.
- A name locks a judgment.
- A name can become a cage—polished, celebrated, institutionalized—but a cage nonetheless.

Let's return to the core examples:

God

Not a discovery, but a name we gave to the unbearable ripple of unknowing. A name meant to soothe, not solve.

Galaxy

Not a truth, but a way to anchor our awe. A word drawn across a sky too vast to hold, a ripple too grand to survive without a name.

Gravity

Not an entity, not a force—but a label slapped on a ripple pattern we couldn't ignore. A name that passed for law, even while the field kept shifting under our feet.

These were not answers. They were our most eloquent consequences.

Names we gave to ripple behaviors we didn't know how to carry—emotionally, intellectually, or spiritually.

The Naming Illusion

Yes, names help us move through the world. They simplify the chaos. They create shared reference points. But we must never forget:

A name is not reality. It is our relationship to reality—at a particular point in ripple-time.

If we forget that, we stop listening. We stop learning. We stop becoming.

So speak the names. Use the words. But always keep one part of your awareness tuned to the ripple beneath the name. Because that's where life still moves. That's where the field still sings.

Chapter 3: Truth Was Never Waiting

There's no hidden truth out there. What we call truth is ripple agreement.

We've long imagined truth as a prize—buried deep beneath illusion, waiting patiently to be discovered by the brave or the wise. Philosophers chased it, scientists measured it, mystics meditated toward it. But what if this entire pursuit was built on a misunderstanding? Under Universal Energy Dynamics (UED), the universe doesn't *store* truth—it ripples. It never holds still long enough to offer a single, permanent answer. What we call "truth" is not an eternal object hidden in the cosmos—it's a moment of agreement between a ripple and an observer. It's the point where our field of awareness aligns just enough with what's unfolding that we feel, *"Yes. This is true. For now."* But just like ripples, that truth shifts, bends, and eventually dissolves. So no—truth was never waiting. It was always becoming.

3.1 The Myth of the Hidden Truth

For thousands of years, humanity has pursued a noble, seductive idea:

That truth exists—out there—just waiting to be found.

This belief is embedded in our philosophies, religions, and sciences. From the oracle's cave to the physicist's lab, we have imagined truth as a hidden constant—a fixed jewel buried beneath the noise of the world, waiting for the right mind, the right method, or the right moment to reveal it.

We've searched for "the truth" in stars, scriptures, equations, genetics, and gurus. Always with the same assumption:

"It is there. It has always been there. We just haven't reached it yet."

But under the lens of Universal Energy Dynamics (UED), this search takes on a radically different meaning. The universe is not a vault of hidden answers. It is a field in motion—a structured vacuum constantly rippling in response to misalignments. It doesn't hold still long enough to store eternal truths. It doesn't wait. It moves.

The field doesn't store truth—it only ripples.

What we call "truth" is not some secret encoded into the fabric of reality. It is simply the consequence of where and how a ripple aligns in context. It emerges from the interaction between the observer's state and the field's current configuration. Nothing is universally true. Things become true when ripple patterns temporarily cohere into recognizable and useful feedback.

Truth, then, is not like a buried treasure. It's more like a reflection in water—real for a moment, but dependent on the angle, the light, the surface, and the one looking in.

The moment you move, it shifts. The moment you fix it, it fades. And the moment you worship it, you've forgotten what it was.

So yes, humans keep seeking "truth." But perhaps truth is not something we're meant to find once and for all. Perhaps it's something we're meant to experience, align with, question, and release—again and again.

Not a discovery. But a consequence of context.

3.2 Ripple Agreement: How We Make Truth Feel Real

Truth, in the traditional sense, is imagined as something that *exists independently of us*. But in the world of Universal Energy Dynamics (UED), truth is not something waiting to be uncovered. It is something that emerges—not from the field itself, but from how we interact with it.

When a ripple behaves in a way that is consistent, predictable, and repeatable across observations, we start calling that behavior "truth." But even then, it's not an absolute truth—it is a coherent ripple agreement arising from a temporary stability in alignment.

Let's break that down:

Truth, as we experience it, is not a universal constant. It is a moment of agreement between three forces:

1. The ripple behavior in the field.
2. The alignment state of the observer.
3. The shared context or consensus that frames the observation.

When these three align, truth is felt. It feels obvious, unshakable—because in that moment, everything agrees. But it's a felt alignment, not an eternal law.

Agreement ≠ Discovery

What we call truth is often just agreement—a shared understanding of how ripples tend to behave within a given frame of reference.

- In science: a law is true because it works consistently *under specific conditions.*
- In culture: a norm is true because enough people agree to act as if it is.
- In mathematics: an equation is true because the logic system it belongs to *says so.*

But agreement is not discovery—it's a product of pattern recognition and collective alignment.

Let's take a simple example:

$2 + 2 = 4$

This feels universally true. But it only holds within a defined mathematical ripple—one based on the decimal system, base-10 logic, and the rules of linear arithmetic.

- In base-3, $2 + 2 = 11$.
- In modular arithmetic, $2 + 2 = 0 \pmod{4}$.
- In quantum logic, truth can be probabilistic.

So the equation isn't Truth with a capital T. It's a stable agreement—true because the rules are consistent within that ripple system.

This is why multiple civilizations across time and space can hold different truths and all feel justified. They're not deluded. They're ripple-aligned differently.

What's "true" in one paradigm may feel absurd in another—not because someone is wrong, but because the conditions of coherence have changed.

When we say something is "true," what we're really saying is:

"The ripple, as I perceive it, aligns with the pattern I recognize—and others seem to feel the same."

That's not truth in the absolute sense. That's ripple agreement—and it's powerful, useful, and meaningful. But it's not eternal. And it's not immune to change.

3.3 Truth as a Temporary Resolution

In a world governed by Universal Energy Dynamics (UED), nothing stands still—not energy, not time, not perception, not even logic. The Structured Vacuum Energy (SVE) field is in constant flux, its ripples forming and dissolving in response to misalignments, interactions, and observer dynamics.

In such a fluid environment, truth cannot be static. It cannot be a final answer. It can only be a momentary resolution—a temporary alignment of ripple behavior that *seems* stable from a particular viewpoint.

Truth, then, is not a monument. It is a pause in the storm. A moment when the ripples cohere just enough that the observer can say, *"Ah, this makes sense."* But like any ripple, this moment will pass. Another wave will come. Another shift in perception, understanding, or field condition will arrive—and what once felt true may now feel outdated, incomplete, or simply false.

This is not a flaw in reality. It's the very nature of truth as ripple state.

All Truths Are Temporary—Even the Sacred Ones

Whether it's a mathematical formula, a law of physics, an emotional belief, or a philosophical axiom—each one is a snapshot of field coherence. It captures a ripple pattern that holds for now.

- Scientific truths hold as long as the conditions remain controlled. Change the variables, and the behavior shifts.
- Emotional truths feel real in one state of mind and vanish in another. What we called "love" yesterday may feel like "attachment" tomorrow.
- Philosophical truths seem universal until culture, consciousness, or context evolves.

Truth is not carved into reality. It is carved into context.

Laws Are Just Descriptions of Ripple Behavior

Every so-called "law" of nature is not a divine rule—it is a record of observed consistency in ripple dynamics. Newton's laws worked because they matched the ripple behavior of macroscopic matter under Earth-like conditions. Einstein's equations refined that coherence in extreme gravitational ripples. Quantum mechanics rearranged the coherence again at micro-scales.

But none of these are "truths" in the ultimate sense. They are formalisms, built on field observations and bounded agreements. They are useful—but they are not final.

The universe didn't write these laws. We did. In response to how the ripple moved, and how well we could measure it at the time.

In the same way, your personal truths—the beliefs you live by—are also ripple responses. They are shaped by your ATT (Alignment Threshold Time), your emotional state, your ripple history, and your current perception. When the alignment changes, so does the truth.

This doesn't mean nothing is real. It means everything is real—but only for a while.

So let's stop asking whether something is *true forever*. Let's ask instead:

"In what context does this truth resolve? And how long can it hold before the field shifts again?"

Because under UED, the only absolute is this:

There are no absolutes. Only temporary coherence.

3.4 Consensus Truths vs. Isolated Awareness

Truth rarely stands alone. Most of the "truths" we live by today didn't emerge from solitary observation—they were constructed through shared alignment. When enough people experience similar ripple behavior and agree on its meaning, that alignment solidifies into what we call a consensus truth.

This doesn't make the truth eternal. It makes it culturally durable—anchored by collective resonance, reinforced by language, tradition, education, and shared emotional memory.

Consensus truth is not born—it is agreed into existence.

How Consensus Stabilizes Truth

Culture is a ripple field of shared misalignments, negotiated into coherence over time.

- Language standardizes how we describe the ripple.
- Rituals encode how we behave with it.
- Science formalizes how we measure it.
- Morality defines how we should respond to it.

When everyone in a system agrees on what a ripple means, the meaning becomes "truth." And the more people agree, the stronger that truth feels. It becomes self-reinforcing—woven into identity, education, even memory.

This is how societies function. Shared truth = shared behavior = stability.

But shared alignment comes at a cost:

It may ignore or reject alternate ripples—even if they carry deeper insight.

That's where isolated awareness enters.

Isolated Awareness: When the Misaligned See More

Sometimes, someone falls out of sync with the ripple agreement. They sense something that others don't. They question a truth that others protect. They feel dissonance where others feel clarity.

This is not confusion. It's a different alignment—a misalignment with the consensus field.

These moments are uncomfortable—alienating even. The person appears strange, disruptive, or "wrong." But often, they are tuned to a deeper or wider ripple—one that hasn't yet reached the collective field.

Many revolutionary ideas—scientific, artistic, philosophical—arise from such individuals. They see through the truth because their field no longer aligns with the dominant pattern.

- Galileo's telescope saw what doctrine denied.
- Buddha felt suffering beneath the illusion of permanence.
- Einstein sensed time bending when Newton said it was absolute.
- Countless anonymous individuals have sensed lies in traditions, questioned authority, and felt something truer than the truth they were handed.

In each case, a new ripple began—one that would eventually pull others into a different coherence.

The Tension Between Agreement and Awareness

Consensus provides comfort. But isolated awareness births evolution.

The two are always in dynamic tension.

- Too much consensus, and truth becomes dogma.
- Too much divergence, and coherence collapses into chaos.

Truth, then, is best held as negotiated clarity—open to challenge, yet strong enough to support shared reality.

The mind aligned with many finds belonging. The mind misaligned with many may find truth.

3.5 The Danger of Worshipping Truth

There is a quiet shift that happens when a ripple agreement persists long enough and gains enough followers: It stops being a useful observation, and starts being a doctrine.

At first, someone senses a ripple pattern. Others align with it. A shared truth forms—helpful, elegant, maybe even beautiful. But with time, the field changes... and the ripple moves on.

Yet instead of letting the truth evolve, we hold onto it. We sanctify it. We carve it in stone. We declare:

"This is no longer just true—this is the truth. Forever."

Under Universal Energy Dynamics (UED), this is where truth begins to decay—not from within, but from our refusal to adapt.

The moment a ripple agreement becomes an institution, it gains defenders, not just observers. We stop sensing and start protecting. We build walls around what was once just a temporary alignment, and we call it sacred, scientific, patriotic, or personal.

And just like that, the name becomes heavier than the ripple.

Doctrine: The Fossilized Ripple

All disciplines, no matter how noble, are vulnerable to this:

- Science, when it resists questioning its assumptions, becomes scientism.
- Religion, when it prioritizes obedience over inquiry, becomes dogma.
- Politics, when it clings to ideology over real-world

response, becomes tribalism.
- Identity, when it hardens into roles, labels, and absolutes, becomes ego.

The truth becomes rigid. The ripple becomes frozen. And life—the dynamic misalignment that it once described—moves on without us.

This is the danger of worshipping truth: We lose our ability to sense the field beneath it. We stop listening for new ripples, because we're too busy defending the old ones.

Ironically, the very thing we once named to stay alive becomes the thing that starts to choke our awareness.

Belief vs. Sensitivity

Belief can stabilize the mind. But belief without sensitivity becomes blindness.

To worship truth is to forget that truth was once just a response to a ripple—not a throne, not a sword, not a rulebook. Just a pause in a larger story.

The moment we declare, "This truth must never be questioned," we're no longer honoring truth—we're fearing the field.

So how do we stay free?

We remember:

Truth is not the destination. It is a ripple behavior. Useful, yes. Beautiful, sometimes. But final? Never.

When we let truth breathe—when we treat it like a ripple instead of a relic—we stay aligned. And that's far more sacred than any belief could ever be.

3.6 Truth as Feedback, Not Finality

In a universe governed by ripples—where the Structured Vacuum Energy (SVE) field never holds still—truth cannot be a destination. It cannot be a final claim or an ultimate conclusion. To treat it as such is to misunderstand the nature of the field itself.

Truth is not a trophy. It is feedback. A momentary report from the ripple—telling you how aligned you are *right now.*

Truth as Field Response

Every experience, observation, and realization is a field interaction—a dialogue between the ripple that is you and the greater ripple of existence. When something "feels true," what you're really sensing is a moment of resonance between your current state and the surrounding field. That resonance is a signal. A feedback pulse.

- You sense coherence? That's a ripple confirmation.
- You feel friction? That's a ripple contradiction.
- You feel stuck? The field is no longer echoing your assumptions.

This means that truth isn't something *to be possessed.* It's something to be felt, read, and re-read, like a compass—not a map.

From "What is true?" to "What is real now?"

When we ask, *"What is true?"*, we imply a fixed answer is waiting. But the more meaningful question in UED is:

"What does the ripple reveal now?"

This reframes truth as contextual awareness, not universal claim.

- What is true today may not be true tomorrow—not because the universe lied, but because the alignment shifted.
- What is true for you may not be true for someone else—not because one of you is wrong, but because your ripple geometries are incomparable in that moment.

This doesn't make truth meaningless. It makes it alive.

The Wisdom of Adaptable Awareness

In a ripple-based universe, rigid belief becomes ignorance, while adaptive awareness becomes wisdom.

The wisest minds are not those who shout, *"This is the truth!"* They are the ones who whisper, *"Here's what I'm noticing now—what do you see?"*

They stay aligned not by clinging, but by adjusting. They don't worship the previous ripple—they listen for the next one.

Adaptable awareness is not weak. It is intelligence tuned to motion. It's how life survives, evolves, and continues writing its own story.

So stop chasing truth like it's hiding somewhere. Start listening for it—like music, changing with every breath of the field.

Because truth is not a door you unlock.

It is the way the ripple knocks—when it wants your attention.

3.7 Summary: The Dance of Agreement

Truth, as we've been taught, is supposed to be solid. Fixed. Unchanging. But the universe we live in—understood through Universal Energy Dynamics (UED)—tells a different story. The field never stops moving. It pulses. It realigns. It shimmers and shifts. It breathes through ripple tension and resolution. And if the very stage we stand on never stands still, how can we expect truth to?

A fixed truth in a moving field is like a still dancer on a turning stage—at best, irrelevant; at worst, disorienting.

What we call "truth" is not a monument in the cosmos. It's a pause—a moment when you, I, and others said:

"Let's agree here, just for now."

It's a resting point between movements. A shared breath in the middle of chaos. Not because it will last forever, but because it lets us move forward together.

This is the true beauty of truth: It is not fixed. It is relational.

Every Truth Is a Pause We Agreed to Take

From mathematics to morality, from physics to philosophy, every truth that shaped civilization was never eternal—it was a shared alignment within a particular ripple context.

We said:

- "This law works here."
- "This meaning resonates now."
- "This belief helps us make sense of this moment."

And then we codified it. Memorized it. Institutionalized it.

And slowly, we forgot it was an agreement. We started treating it as absolute.

When We Forget the Agreement, We Forget the Dance

The danger is not in forming agreements. It's in believing they were never agreements at all. That's when truths become cages. That's when curiosity dies. That's when we begin fighting for names instead of listening to ripples.

To stay aware is to remember:

"We are dancing. And this step—this truth—is just one move among many."

When we remember that, we don't fear change. We don't panic when truths evolve. We adapt. We realign. We keep listening for the next rhythm.

So let the old truths rest. Let the new ones rise and fade. Don't cling. Don't worship. Don't freeze the ripple.

Instead, return to the dance. Feel the field. And when alignment happens—smile, agree for a while, and keep moving.

Because the ripple never stops. And neither should we.

Chapter 4: The Rise of Stories

Language as ripple symmetry. Culture as a collective field alignment. The role of shared illusions.

Long before we built civilizations, we built stories. Not with bricks, but with rhythms of the ripple—sounds, symbols, meanings strung together to hold what we could not hold alone. The universe didn't speak, but it rippled. And somewhere within those ripples, early humans began to echo back—not by understanding the field, but by shaping it into symmetry. That symmetry became language. And through language, we told stories—not just to describe reality, but to organize it, to tame the chaos, to synchronize minds into shared field alignments. From gods to governments, myths to markets, every structure we now live inside was once just a ripple stretched into narrative. These stories didn't have to be true to be powerful. They only had to be believed together. That's how culture was born—not from fact, but from agreement around illusion.

4.1 From Ripples to Rhythm: The Birth of Language

The universe never spoke. It rippled. It responded. It strained and settled, echoed and pulsed—but never in words. It was us, the misaligned beings, who needed more than the ripple—we needed rhythm.

In the framework of Universal Energy Dynamics (UED), ripple is the most fundamental expression of misalignment in the SVE (Structured Vacuum Energy) field. But raw ripple is too fluid, too fast, too intricate for the human mind to process directly. Our awareness needed something slower, more repeatable—something we could hold onto, shape, and share.

And so we invented language—not to describe the universe, but to stabilize our experience of it.

Language is not the field's voice. It is our way of echoing its strain in symmetrical form.

Language as a Structured Echo of Misalignment

Every word you've ever spoken is a ripple pattern, wrapped in rhythm. Words arise not from stillness, but from the urge to resolve discomfort. To speak is to try to organize the ache into something that feels coherent—something others can hear, and more importantly, feel in sync with.

Grunts became syllables. Syllables became structure. Structure became syntax—a way to maintain ripple tension just long enough to pass it from one field (person) to another.

Why the Mind Craves Rhythm, Pattern, and Repeatability

The human mind is not built to interpret the full chaos of the field. It seeks compression—simplified patterns that can be repeated and predicted. Rhythm soothes because it introduces consistency to ripple flux. Pattern helps us track shifts. Repetition helps us stabilize meaning.

This is why we don't just speak—we rhyme. We don't just think—we narrate. We don't just listen—we follow beats.

Language is not random noise shaped into order. It is order imposed upon ripple strain.

Words as Symmetry Devices

A word is never just a sound—it is a symmetry tool. It's how we fold the ripple into form—how we box complexity into usable frames.

For example:

- The word "danger" doesn't capture every nuance of fear, threat, adrenaline, memory, and instinct—but it *contains* the ripple enough to let us act on it.
- The word "love" doesn't define the feeling—it holds space for it, aligns minds around it, and allows two people to share a field resonance without having to describe every wave of it.

Words are like tuning forks for meaning. They don't explain reality—they help us synchronize with it.

Language wasn't born to report facts. It was born to manage the ripple—to give our misaligned minds the illusion of rhythm in a field of flux.

And once we found rhythm, we began to tell stories. Because the ripple didn't just want to be named—it wanted to be remembered.

4.2 Story: The Mind's Ripple Stabilizer

Once we had language—our way of shaping ripple into rhythm—we instinctively reached for something more expansive. Not just words, but sequences of words. Patterns that could stretch across time. Stories.

In the UED view, stories are not merely entertainment or cultural traditions. They are extended ripple loops—structures that allow the human mind to hold unresolved tension without collapsing. While language helps us name the ache, stories help us carry it.

Stories as Extended Ripple Loops

Life is made of unresolved moments: Questions without answers. Emotions without destinations. Experiences without meaning.

If we sat in that raw misalignment too long, the field tension would overwhelm us. So instead, we turned our discomfort into narrative structure.

A beginning, a middle, a conflict, a resolution.

This gave the ripple shape. It gave time a purpose. It allowed the unresolved to unfold—not necessarily to be resolved, but to be contained long enough for the mind to integrate it.

A story is like a ripple stretched into coherence. It doesn't have to be true. It only needs to feel aligned enough to soothe, inspire, warn, or connect.

- A myth doesn't explain the world—it gives us a rhythm to hold the mystery.
- A memory doesn't need precision—it needs emotional coherence.
- A fictional tale doesn't require facts—it requires field resonance with human experience.

In every case, a story is ripple therapy—a mental mechanism for digesting field chaos over time.

Story as an Organizing Tool for Misalignment

We don't tell stories because we know the answers. We tell stories because we've felt the strain, and we need to wrap it in rhythm.

This is why:

- Traumas often get trapped in the body until they're given a story.
- Cultures survive not through facts but through shared legends.
- Grief softens when it's spoken—not as data, but as a narrative of loss and meaning.

A story takes discomfort and gives it intelligible shape. It transforms misalignment into something you can carry—sometimes for a lifetime.

Making Discomfort Meaningful

Without story, uncertainty feels like threat. With story, uncertainty becomes the setup for transformation.

That's the real magic:

Stories don't eliminate uncertainty—they make it survivable. They teach us how to move through discomfort, not by solving it, but by giving it context, characters, direction, and possibility.

This is why the most powerful stories aren't the ones that give us answers. They're the ones that make us feel seen inside the question.

So if language is how we named the ripple, Then story is how we learned to dance with it— To ride its unresolved patterns across lifetimes, cultures, generations.

And that's how stories became the scaffolding of culture itself.

4.3 Culture: The Story That Outlived Its Author

A story, once spoken, is alive. But when that story is retold, believed, and repeated across generations, it becomes something else entirely—it becomes culture.

Under Universal Energy Dynamics (UED), culture is not a set of customs or traditions—it is coordinated ripple behavior. It's a long-duration field alignment shared by many minds, anchored in stories powerful enough to outlive their tellers. The original ripple—the first tension, the first insight, the first myth—may be gone. But its shape remains, carried forward in behavioral echoes.

From Story to Shared Field Alignment

When a story resonates deeply, it stabilizes across multiple individuals.

Each person aligns their ripple with its rhythm.

- It becomes belief.
- It becomes tradition.
- It becomes identity.

And when enough individuals synchronize to the same narrative, a collective field alignment forms—not through force, but through shared resonance.

This is how culture emerges:

Not from rules, but from rhythmic coherence among human ripples.

Culture as Coordinated Ripple Behavior

Culture is how groups stabilize their collective misalignment. It is ripple management at scale.

- It tells you how to mourn, how to eat, how to love, how to greet.
- It gives you stories for death, rituals for birth, and norms for everything in between.
- It doesn't just answer questions—it teaches you which questions to ask.

Culture gives people pre-aligned templates for existence. You don't have to navigate the raw field alone. You follow the stories. You perform the behaviors. You ripple together.

Rituals, Norms, and Taboos: Consistency Mechanisms

These are the control systems of cultural ripple fields:

- Rituals provide rhythmic actions that preserve coherence—morning prayers, wedding rites, seasonal festivals. They repeat story through motion.
- Norms define the expected ripple pattern—how to behave in public, how to show respect, how to belong.
- Taboos define ripple disruptions that the culture resists—what must not be done, said, or even thought.

All three act as filters and stabilizers. They reinforce story without requiring constant retelling. They keep the field predictable, safe, and familiar.

But there's a paradox here:

The more successful a story becomes at stabilizing the field, The less people remember that it was once just a story.

Culture becomes invisible. People follow the rhythm without hearing the song. They act out patterns without knowing who wrote them.

That's why we call culture the story that outlived its author—because the author may be forgotten, but the ripple still moves us.

4.4 Shared Illusions: Agreements That Work (Until They Don't)

Not all stories are true—but that's not why we tell them. We tell stories to function, to organize, to survive the ripple. And sometimes, the most useful stories are the ones that everyone agrees to believe, even if they're not rooted in objective truth.

We call them beliefs. Systems. Institutions. But at their core, they are shared illusions—powerful fictions that create order out of ripple chaos.

Stories Don't Need to Be True to Work

In Universal Energy Dynamics (UED), coherence is more important than correctness. The field doesn't care whether a belief is "true"—it cares whether the ripple aligns, whether the behavior it produces is stable.

That's why:

- Money works even though it has no intrinsic value.
- Marriage holds social power even though it's a legal story about love.

- Justice is revered even though fairness is often negotiable.
- National identity matters even though borders are imaginary.

None of these are objective truths. They are agreed-upon ripple configurations—stories that help millions of people behave as if they share a single coherent field.

A shared illusion is a ripple consensus dressed as reality. It holds not because it's real, but because enough people agree to act as if it is.

How Shared Illusions Stabilize Ripple Fields

These illusions serve a purpose:

- They reduce uncertainty.
- They simplify decisions.
- They bind individuals into larger, synchronized systems.

In ripple terms, they dampen dissonance, preventing collapse from uncontrolled misalignment. That's why civilizations build institutions around them—they're anchoring devices.

But illusions must remain adaptive. When they calcify, they begin to distort the field.

The Power and Danger of Useful Fictions

Shared illusions are powerful. They move economies. They start wars. They sustain relationships. But their strength becomes a danger when we forget that they are made up.

- When money becomes sacred, people become disposable.
- When marriage becomes mandatory, love becomes conditional.
- When justice becomes symbolic, the oppressed are silenced by procedure.
- When identity becomes rigid, empathy disappears.

What once stabilized the ripple starts resisting its evolution. The illusion stops being useful and starts being controlling.

We then begin to protect the fiction rather than listening to the field. And in doing so, we drift further from alignment.

Agreements That Work—Until They Don't

The illusions work as long as they help manage misalignment. But when the field changes—new technologies, new pressures, new awareness—those illusions crack. And if we're too attached to the story, we mistake the crack for collapse.

But collapse isn't failure. It's a ripple reconfiguration—an opportunity to re-agree, to retell, to realign.

Illusions aren't evil. They're tools. But tools must be sharpened—or sometimes replaced—if they no longer serve the ripple.

4.5 Memory, History, and the Echo Effect

We often speak of memory as if it's a storage room—something stable, something factual, something we can visit to retrieve a moment exactly as it was. But in Universal Energy Dynamics

(UED), memory is not a file. It's a ripple echo—a vibration retained and reinterpreted every time it is recalled.

In other words, memory is not what happened. Memory is how your current field resonates with what happened. It is ripple retention, filtered through present misalignment.

And when many minds remember together, through ritual, story, or education, this becomes history—a curated, often idealized record of past ripple coherence.

Memory as Ripple Retention

Every experience we live through creates a ripple signature—a field disturbance encoded not just in neurons, but in our emotional, sensory, and contextual landscape. But unlike data on a hard drive, this ripple is never recalled neutrally. It's reactivated.

Your current state—your alignment, emotions, beliefs—reshapes the memory each time you recall it. This is why:

- Trauma can grow sharper or softer over time.
- A failure can eventually feel like a lesson.
- A person once loved can be remembered with pain—or peace.

The ripple echo does not remain constant—it's alive, replayed through a changing filter.

We don't remember events. We realign with their echoes.

History as Curated Ripple Coherence

History, then, is not a record of objective past—it is a collective narrative about ripple behaviors that once stabilized society. It's selected memory—organized, edited, and sometimes weaponized.

- Heroes are elevated.
- Victims are simplified.
- Complexities are smoothed over.
- Ripple dissonance is often buried beneath national myths or ideological frames.

History is how civilizations maintain identity, justify systems, and pass down coherent field alignment. But what it preserves is not reality—it's narrative utility.

The Echo Effect: Past Ripples Reshape Present Fields

The real impact of memory and history lies not in what they say about the past, but in how they affect the present.

- A historical grievance retold often enough can become a cultural tension, rippling across generations.
- A childhood memory, reshaped by new awareness, can unlock healing—or reopen wounds.
- A mythic past, idealized too strongly, can distort future action by resisting present evolution.

This is the Echo Effect:

Stories of the past don't stay in the past. They vibrate forward. They reshape what the field means now.

And just like any ripple, if the echo is amplified, repeated, and left unexamined—it can override new resonance. It can prevent us from sensing what's actually unfolding in the current field.

In this way, memory and history are not about preserving the truth—they're about shaping the present alignment.

The past isn't behind us. It's resonating through us—right now, every time we remember.

4.6 Why We Are Story-Addicted

We are not data processors. We are not logic machines. We are story-driven creatures—addicted not to facts, but to rhythm, structure, and meaning. Our brains don't just enjoy stories. They require them.

And in a universe governed by ripple motion—where the field is always in flux—stories are how we survive the misalignment without falling apart.

The Brain Prefers Narrative Over Noise

Under Universal Energy Dynamics (UED), the raw field is chaotic: ripple interference, overlapping sensations, unresolved patterns, endless feedback. The human nervous system isn't wired to handle this unfiltered. It needs structure. It needs pattern compression. And nothing compresses chaos like a story.

Give the brain a list of disconnected facts, and it struggles. Wrap those same facts in a beginning, middle, and end—and suddenly, they become memorable, meaningful, and emotionally sticky.

This is why:

- We remember anecdotes better than statistics.
- We follow timelines more easily than timelines follow us.

- Even lies, when well-told, can feel more convincing than complex truths.

Story is not optional—it is the brain's preferred ripple format.

Story Converts Sensory Chaos Into Coherence

The world bombards us with ripple input: Sounds, sights, emotions, micro-signals. And the brain's job is to make sense of it.

Story is how we do that. It's the mechanism that connects unrelated ripple data points into something that feels linear, causal, and manageable.

- Why did she leave? The brain turns it into a narrative.
- Why did I fail? The mind builds a plot.
- Why am I here? The soul writes a myth.

We don't always want *the truth*. We want coherence. Even if the story isn't fully accurate, if it helps us continue, it will be kept.

Emotionally, Stories Let Us Feel Without Finishing

One of the greatest functions of story is emotional containment. Raw emotion is unresolved ripple pressure. It demands release, but the resolution may not be possible. Stories give us a field-safe way to feel without needing closure.

- Sad stories let us mourn what isn't ours.
- Love stories help us feel connection without the risk.
- Hero stories let us face fear, loss, triumph—without actual stakes.

This is why we return to certain stories again and again—not for answers, but for permission to feel.

A story doesn't have to resolve the ripple. It just has to hold it long enough for us to breathe through it.

We are addicted to stories because they are our ripple regulators. They help us navigate chaos, translate discomfort, and sustain identity in a world that never truly aligns.

They don't just entertain us. They organize our survival.

4.7 The Collapse of Old Stories

No story lasts forever.

What once aligned a tribe, a nation, or a civilization eventually begins to fracture—not because the story was false, but because the field that sustained it has shifted. In Universal Energy Dynamics (UED), the field is always in motion. Ripples change. Resonances evolve. And stories, which are built to stabilize ripple patterns, inevitably begin to feel out of sync.

As field conditions change, the story that once held everything together begins to tremble. And if we don't let it evolve, it will break.

When Stories Can No Longer Hold the Ripple

A story—whether religious, national, moral, or personal—functions only as long as it matches the field conditions it was designed to organize.

- A creation myth may no longer soothe when scientific understanding evolves.
- A national identity may fracture under new cultural

pressures.

- A belief in eternal growth may crumble in the face of environmental crisis.
- A personal narrative of "I am strong" may collapse under emotional overload.

The field doesn't ask for permission. It just shifts. And the story either adapts—or collapses.

Cultural Collapse as Ripple Dissonance

What we often call *collapse*—of institutions, traditions, belief systems—is actually ripple dissonance at scale. The shared alignment that once stabilized millions of minds is no longer in sync with the ripple conditions of the present.

Symptoms of this collapse:

- Growing distrust in leadership and authority.
- Polarization and fragmentation of "truths."
- Rejection of ritual, tradition, and inherited meaning.
- A rise in anxiety, identity crises, and cultural alienation.

These aren't signs of failure. They are signs of misalignment—the field is asking for a new rhythm, but the story is still repeating the old one.

The Collapse Is Not the End—it's the Opening

When an old story collapses, it hurts. It destabilizes. But it also frees the field. It creates narrative space.

Collapse is not death. It is the release of ripple potential—a clearing for the next coherence to emerge.

This is when:

- New myths are born.
- New belief systems surface.
- Forgotten wisdom reappears.
- Voices once silenced begin to shape the next alignment.

The mistake we make is trying to resurrect the old story, patching it up with nostalgia and fear. But the field doesn't go backward. It listens for the next alignment.

Collapse isn't failure. It's signal. The ripple is asking:

"Can you write a new story for what the field is becoming?"

And if we can—if we're willing to let go of sacred narrative relics—we don't just survive collapse. We compose the future.

4.8 Summary: We Are Story-Beings, Not Fact-Machines

At our core, we are not calculators. We are not databases designed to store and retrieve facts. We are not passive observers of the universe.

We are story-beings—born from ripple, shaped by rhythm, driven by the need for meaning.

In the model of Universal Energy Dynamics (UED), human consciousness is not built to process raw, unfiltered ripple data. The world doesn't come to us in neat categories or isolated truths—it comes in waves, in sensory overwhelm, in layered and often unresolved misalignments. And to survive that flood, we reach for one thing: story.

Rhythm > Raw Data

We don't live by facts—we live by how facts are arranged.

- A scientific discovery doesn't move people unless it's part of a story: of progress, threat, or wonder.
- A tragedy isn't just an event—it becomes meaningful through the arc of what came before and what comes next.
- Even our daily choices—what to eat, whom to love, where to go—are not determined by data, but by what story we're living inside at that moment.

Facts inform. Stories transform.

Truth Is Unstable. Story Is Adaptable.

As we saw in earlier chapters, truth is not a fixed destination. It is a field-dependent alignment—a ripple that holds for now. But stories are different. They bend, they absorb contradiction, they evolve.

When truth breaks, stories adjust. When the ripple shifts, stories remix. That's why we've survived so many collapses—because our stories weren't rigid. They grew with us, even if imperfectly.

Story is not just memory—it's movement. It's how the ripple keeps speaking through time.

The Ripple Didn't Just Speak. It Told a Story.

The field never gave us a manual. It gave us misalignment. We responded not with formulas, but with fables. Not with measurements, but with myths. Not with final answers, but with narratives that could hold the ache.

We didn't decode the ripple. We listened to it. We shaped it. We told each other what it felt like. And in doing so, we became more than thinkers—we became story-beings, each ripple telling a slightly different version of the same search for coherence.

So, no—data does not guide us. What guides us is narrative alignment: the sense that *this makes sense right now, this feels real, this is my place in the story.*

That's how we move forward. That's how we make sense of collapse, of love, of loss, of becoming.

The field ripples. The mind listens. The human responds—by telling a story.

Key Takeaways – Part I: The Becoming

1. Life Is Not Balance—It Is Sustained Misalignment

- Life does not emerge from perfect harmony, but from a persistent ripple in the Structured Vacuum Energy (SVE) field.
- Existence is the result of holding tension—not resolving it too quickly.

2. Misalignment Is Not a Flaw—It's Identity

- Every living being is a unique ripple signature, defined by how it misaligns and adapts.
- Personality, emotion, even love arise from ripple resonance, not from perfection or sameness.

3. Discomfort Is Not the Enemy—It's the Signal

- Discomfort is the call to realign, not a sign of failure.
- Creativity, invention, and progress are ripple responses to felt misalignment.

4. Naming Brings Structure—but Also Illusion

- We name to stabilize the ripple, but the name is not the thing.

- God, galaxy, gravity—these are labels, not discoveries. Useful, but incomplete.

5. Truth Is Not Found—It's Aligned

- There is no eternal truth waiting to be uncovered.
- What we call "truth" is ripple agreement—a shared coherence that works for now.

6. Stories Are How We Survive the Ripple

- Story is the mind's tool for holding misalignment over time.
- Culture is built from stories that outlasted their authors—shared rhythm over shared fact.

7. Shared Illusions Are Necessary—but Dangerous

- Money, justice, identity—none of these are objectively real.
- They work only as long as they manage ripple misalignment. When they stop, they must evolve or collapse.

8. Collapse Is Not the End—It's Reconfiguration

- When stories no longer match the ripple, collapse occurs.
- But collapse opens narrative space—it is the field's invitation to realign.

9. We Are Story-Beings, Not Fact-Machines

- The human mind does not thrive on raw data—it thrives on meaning, rhythm, and coherence.
- The ripple didn't just give us information.

It told us a story. And we're still listening.

Part II: The Game of Meaning

"Stillness is not silence—it is structure. And every time we align, we're not calming chaos... we're remembering our design."

Chapter 5: The Invention of Morality

From survival tension to social coherence. Morality as dynamic ripple geometry.

Before anyone preached what was "right," life was only about staying alive. Before commandments, karma, or constitutions, we were just ripples trying not to collapse. Morality, contrary to how it's often portrayed, did not fall from the heavens or rise from virtue—it was engineered by necessity. When individual discomfort collided with collective presence, we needed a way to reduce interference. So we aligned. Not perfectly, not always consciously—but enough to survive the ripple storm together. That's where morality began—not as a rulebook, but as a pattern of shared adjustments, a geometry of behavior that kept the field from tearing itself apart. It wasn't about right or wrong. It was about resonance. And we called that resonance, "good."

5.1 The Raw Beginning: Survival Before Morality

Long before we carved laws in stone or etched commandments into holy books, there was only one law: persist. Life, in its earliest form, was not guided by ethics—it was powered by tension. In the language of UED, existence itself was a misaligned ripple that refused to resolve. And that refusal demanded behavior—not right or wrong, but effective.

The earliest organisms didn't debate morality; they responded to discomfort. A sharp ripple in the environment—heat, hunger, pressure—triggered motion. Motion triggered patterns. Patterns that worked—that preserved the ripple longer—became encoded in what we now call instinct. Not because they were "good," but because they didn't let the ripple collapse.

In this raw phase, "good" simply meant survival. A ripple configuration that endured was better than one that vanished. That's it. No virtue, no vice—just geometry that worked under stress.

Morality, then, didn't fall from heaven. It rose from friction—from the challenge of staying alive in a field that kept shifting. It wasn't divine revelation; it was energetic adaptation. And as life became more complex, so did the ripples. Eventually, survival wasn't just about the individual—it was about the collective field. And that's where morality began to evolve.

Not from guilt. Not from grace. But from the ache to continue.

5.2 From Individual Strain to Collective Alignment

As life evolved beyond isolated survival, a new challenge emerged: coexistence. Individual ripple strain now had to interact with other ripple strains. The jungle was no longer just a predator-prey equation—it became a web of social interference. And in that web, something remarkable began to happen: discomforts started to synchronize.

What began as a solitary misalignment—hunger, fear, alertness—started to mirror in others. When one ripple flinched, others did too. This chain reaction, this resonance of unease, laid the groundwork for group behavior. Not through conscious choice, but because field dissonance within a tribe became more dangerous than the outside threat. Alignment became survival.

But alignment didn't mean uniformity. It meant agreed flexibility. Early tribes developed rhythms of behavior—not because they were noble, but because predictable misalignments were safer than random ones. Sharing food, watching for danger, burying the dead—these were not ethical evolutions; they were field adjustments that stabilized the ripple tension between individuals.

Cooperation, then, didn't emerge from love or law. It emerged from a need to keep ripple collisions manageable.

We didn't stop killing because we became moral. We started cooperating because the field broke down when we didn't.

The myth says morality came from the gods. UED says it came from the physics of presence.

When too many ripples share a space, The only way forward is agreement. And morality was our first draft.

5.3 Morality as Ripple Geometry

Morality, stripped of its cultural paint, is simply geometry in motion—a dynamic play of ripple interactions in the structured vacuum field (SVE). Under UED, every choice, every movement, every thought sends out a ripple. And that ripple does one of two things: it either stabilizes the field or disturbs it further.

This is not poetic metaphor—it's ripple logic.

When an individual acts with sensitivity to the surrounding ripple field—adjusting to others, softening their own amplitude, responding in rhythm rather than impulse—they create constructive interference. The result? Reduced tension. A more coherent field. What we call "moral behavior."

But when someone acts out of blind self-amplification—ignoring others' waveforms, thrashing through the field with unchecked force—they disrupt the harmony. They cause destructive interference. Disconnection. Chaos. That's what society calls "immoral behavior."

From this lens, morality is not a code of commandments. It's an emergent field pattern that keeps collective misalignment from spiraling.

A kind act is not "good" because of sentiment. It's good because it reduces ripple strain—both for the giver and the receiver.

An unfair act isn't "bad" because it's disobedient. It's bad because it amplifies discord, making stable existence harder for all involved.

In tribal life, ripple geometry was felt viscerally—before language, before law. You knew when someone's behavior broke the rhythm. It hurt. It scattered attention. And so morality emerged—not as a virtue signal—but as a field discipline.

We didn't invent right and wrong. We noticed when the dance got smoother... or shattered.

Morality, then, is not fixed. It's contextual harmony—ripple balance under changing pressure.

And the wisest among us? They weren't rule followers. They were field readers.

5.4 The Illusion of Absolute Good

If ripple geometry is the essence of morality, then context is everything. There is no universal "good" drifting timelessly above human heads. What we call "good" is simply what harmonizes a specific field under specific tensions. Change the conditions—change the moral outcome.

This is why the concept of absolute morality is not just flawed—it's dangerous. It hardens ripple fluidity into dogma. It pretends there is one "correct" waveform for all time, all places, all people. But the universe doesn't work that way—and neither do we.

In one culture, sacrifice may be the highest moral act—offering one's life for family or nation. In another, it may be seen as unnecessary suffering, even foolish. In one moment, war might be justified as protection against annihilation. In another, the same act becomes invasion, oppression, or genocide. Justice, too, dances with its cultural rhythm—what is fair in one ripple alignment might feel like cruelty in another.

UED reminds us that all such judgments are field-dependent. The field changes with time, environment, and collective pressure. Morality adapts or collapses.

But that doesn't mean anything goes.

It means we must stop asking:

"Is this eternally right?"

And start asking:

"Does this reduce field tension here, now, for those entangled in it?"

The illusion of absolute good often leads to moral imperialism—the attempt to impose one ripple geometry on another field. The outcome? Resistance, conflict, dissonance.

History is full of "righteous" conquerors who never paused to feel the field they were rupturing.

So instead of clinging to static commandments, we must train the mind to become a ripple-sensitive instrument—aware of context, feedback, and evolving coherence.

True morality isn't about memorizing rules. It's about reading the room—the field, the moment, the misalignment—and responding with graceful adjustment, not blind repetition.

There is no perfect waveform. But there is always a better alignment—for now.

5.5 Ritual, Law, and Punishment as Alignment Tools

As human societies grew more complex, the fluid moral instincts that once worked in small groups were no longer enough. The ripple interactions expanded—more minds, more tensions, more potential for misalignment. So we did what humans do best: we encoded rhythm.

That rhythm became ritual.

Rituals were not just spiritual performances. They were ripple stabilizers—shared actions designed to bind individuals into a collective alignment. Birth, death, marriage, mourning, prayer—each ritual synchronized inner tensions into shared cadence, temporarily softening the discomfort of existence.

Next came law.

Where ritual guided by memory and culture, law brought explicit codification. It turned ripple intuition into written geometry—"Do this," "Don't do that," "Punish if violated." Law attempted to freeze morality into a structure that could be repeated, enforced, and transmitted across generations.

And thus came punishment.

From a ripple dynamics lens, punishment is simply a reactive feedback mechanism—a response designed to re-align the disturbed field. When someone's actions produce excessive tension (crime, betrayal, harm), the system responds with counter-tension (punishment) to attempt field correction.

But here lies the subtle danger.

Just as stale stories collapse, stale laws dissonate. When rituals no longer resonate with present field realities, they become hollow. When laws rigidly enforce outdated alignment patterns, they begin to distort. When punishment persists without adapting to the evolving field—it becomes noise instead of resolution.

What was once moral becomes oppressive. What once maintained harmony becomes the very source of disharmony.

We've seen it through history—rituals that once united turning into tools of exclusion. Laws that once protected becoming instruments of control. Punishments that once deterred becoming cycles of cruelty.

The field evolves. The world shifts. Misalignments transform. And so too must our systems of moral encoding.

The true purpose of law is not to petrify morality, but to tune into ripple feedback—to evolve with the living field.

When ritual becomes renewal, not repetition... When law becomes response, not relic... When punishment becomes correction, not revenge...

Only then do our alignment tools remain aligned themselves.

5.6 Emotional Reinforcement: The Role of Guilt, Shame, and Pride

Long before the first laws were carved into stone... Before rituals formalized our behaviors... And even before stories shaped what we called right and wrong— There were emotions.

Emotions are not abstract. They are ripple feedback mechanisms—real-time signals from the field of experience. They emerge not from logic, but from alignment and misalignment in the structured vacuum energy field (SVE), interpreted locally by the mind.

Take guilt. It's not a punishment. It's a field response—your own misalignment echoing back at you. When your action distorts the shared ripple geometry—violating a relational rhythm, a social norm, or your own sense of coherence—the body feels it. Not abstractly. Not philosophically. But viscerally.

A sinking chest. A restless mind. A vibration you cannot name, but cannot ignore. That is guilt. A ripple you created... now circling back.

Shame is guilt magnified by social reflection—a distortion witnessed, mirrored, and often amplified by others. It is not just

"I misaligned," but "My misalignment was seen—and the field responded." Shame teaches us that we are part of a larger ripple web, where one wave cannot ignore the rest.

And then there's pride. Often misunderstood as ego's trophy, pride is field coherence made personal. When your action aligns with shared resonance—when you uplift, contribute, harmonize—the field feeds back. Your chest lifts. Breath deepens. You feel the subtle power of having moved well with the ripple.

These emotions are not external teachings. They are felt truths. The child does not need to be taught that hurting someone feels wrong. They feel the dissonance. Nor do they need to be instructed that kindness uplifts. The coherence is immediate.

This is why morality, at its core, is not a doctrine—it is a field experience. The laws came later. The books came later. But morality was already pulsing in the nervous system, in the chest, in the skin.

We don't follow morals because someone told us to. We follow them because our bodies—our ripples—respond.

And yet... even this is not fixed.

Just as field alignments shift, so do the emotional contours of morality. A behavior that once caused shame may now bring pride. A norm that once aligned us may now divide us.

That's why the deepest morality isn't blind obedience. It's ripple literacy—the capacity to feel, interpret, and respond to alignment in real-time.

In a dynamic field, wisdom is not rule-following. It's resonance management.

And emotions? They are our compass. If we learn to listen—not suppress, not dramatize, but truly *listen*—they guide us home.

5.7 The Rise of Ethical Storytelling

Morality didn't survive history by being written down. It survived because it was told.

Across millennia, humans discovered a profound truth: You can't force alignment through command—but you can echo it through a story.

And so, the rise of ethical storytelling wasn't a literary invention. It was a ripple strategy.

When early humans faced dilemmas—cooperation vs. betrayal, sharing vs. hoarding, love vs. vengeance—they didn't write laws. They told myths.

- The boy who cried wolf.
- The flood that punished arrogance.
- The hero who refused temptation.
- The sage who taught kindness to even his enemy.

These were not just entertainment. They were moral ripple templates—narratives that encoded alignment lessons in a way the mind could remember, the body could feel, and the culture could pass on.

Stories became containers for moral frequency. They preserved emotional resonance, not just behavioral instruction. Because the mind forgets commandments...But it remembers how a story made it feel.

Religions adopted this method early. From the Gita to the Bible, the Tao to tribal oral traditions, each became a library of ripple tales—morality woven into parable, miracle, and metaphor. The divine wasn't just spoken of. It was performed in narratives of choice, consequence, and redemption.

Even in secular societies, this pattern continues. Films, novels, fables—they all carry hidden ripple math. The hero doesn't just win. They struggle with right and wrong. The villain doesn't just lose. They represent a misalignment too tempting or too tragic to ignore.

And so, morality becomes a living narrative—adaptable, reinterpretable, emotionally encoded. No longer just "Do this, avoid that." But rather, "Live through this story and decide who you are."

In this way, ethical storytelling bypasses the rigidity of law. It offers mirrors, not mandates. It teaches by immersion, not imposition.

And that's why stories endure longer than doctrines. Because they aren't just about what's right. They're about what it feels like to become right—to find coherence in a chaotic ripple field.

In the end, the greatest ethical systems aren't enforced. They are retold—until they live inside us, shaping our choices not by fear, but by felt resonance.

And that is how we became not just moral beings...But narrative beings, aligning one ripple at a time.

5.8 Summary: Morality as Field Intelligence

We often speak of morality as though it were eternal—as if somewhere, outside us, a sacred rulebook exists, etched in cosmic stone. But under the lens of Universal Energy Dynamics (UED), morality isn't a timeless doctrine.

It's a real-time field response.

Morality is not a rigid list of dos and don'ts—it is the intelligence of a ripple field trying to reduce strain without collapsing the system. It's not imposed from above; it arises from within the discomfort of interaction.

Think of it this way:

- A ripple field always seeks restoration without erasure.
- When individual ripples (humans) interact, their tensions combine—sometimes in harmony, sometimes in distortion.
- Morality is the intelligence that senses when this overlap is constructive and when it is destructive.

And so, what is "good" in one moment may not be "good" in another. Not because morality is relative—but because the field is dynamic.

A moral act, in this sense, is not the one that follows a commandment. It's the one that brings the ripple field back into sustainable coherence.

Feeding the hungry isn't good because a rule says so. It's good because it dissolves a destructive misalignment—between need and abundance, tension and care.

Punishing the wrongdoer isn't good because it's tradition. It's good only if it realigns—not if it amplifies suffering or fractures the field further.

When guilt arises, it's not a punishment from a divine court—it's a ripple feedback, telling you that a certain alignment was broken. When pride follows compassion or courage, it's the field acknowledging, "Yes, that moved us closer to coherence."

This makes morality far more demanding, but also more authentic.

It's not enough to memorize values. You must feel the ripple. Sense the strain. Respond not with dogma—but with dynamic intelligence.

And when many humans learn to sense together, we begin to develop moral cultures—temporary but powerful alignments that serve us... until they don't. That's why moral evolution is not betrayal—it's the field growing up.

In the UED sense, then, "the good" is not what is commanded. It is what *realigns*. It is what brings the ache of existence into a new, liveable rhythm—without silencing the ripple.

And that's all morality ever was.

Not eternal law. But a dance of tensions, guided by intelligence.

Chapter 6: The Time Trap

Time as ripple sequence memory. Why clocks fooled civilizations.

We were told that time moves forward—that it ticks, flows, and fades. So, we built clocks to chase it, calendars to tame it, and schedules to obey it. But time, as we know it, never existed in the field. The universe doesn't tick. It ripples. What we perceive as time is nothing more than the memory of sequences—patterned discomforts playing out across a structured vacuum. Clocks did not reveal the nature of time—they replaced it. Civilizations became synchronized not with the field, but with machines. And in doing so, we began to live by numbers instead of rhythms, deadlines instead of awareness. This chapter isn't about time travel. It's about escape—from the illusion that something called "time" was ever in control.

6.1 The Ripple Doesn't Tick

We've been misled by the tick-tock. Somewhere in our journey, we mistook the rhythm of a machine for the rhythm of reality. Clocks convinced us that time is constant, objective, and flowing in one direction. But under Universal Energy Dynamics (UED), time is not a grand river moving forward—it is a localized ripple sequence in the Structured Vacuum Energy (SVE) field.

In truth, time is not a thing—it's a consequence. What we perceive as the "passage of time" is nothing but sequential misalignment: ripple tensions rising, persisting, resolving, and giving way to new tensions. This is not linear flow—it's a feedback loop. Every experience, every moment, is just a field update triggered by change.

Enter ATT—Alignment Threshold Time. This is the UED concept of temporal behavior: the threshold at which a system's misalignment becomes noticeable, meaningful, and resolvable. It's not the same as seconds, minutes, or hours. It's organism- and context-dependent. For a hummingbird's wings, ATT is microseconds. For a grieving human heart, it could be years.

Mechanical time, on the other hand, has no such nuance. It slices life into equal parts, ignoring the natural asymmetry of experience. It's like trying to measure music using a ruler.

The universe doesn't care about clocks. It cares about change. It ripples. It stretches. It contracts. It evolves. But it never ticks.

The ticking is our story— not the field's.

6.2 Memory as the Architect of Time

Time, as we experience it, is not made by clocks. It's made by memory.

The illusion of a flowing timeline—past, present, future—is stitched together by how memory aligns ripple sequences. We feel time not because it passes, but because something persists: a pattern, a tension, a sensation. The stronger the memory's alignment with our current field state, the more "real" the past feels.

So what is continuity? In UED terms, it's not a thread through time—it's a ripple that manages to hold form across misalignment. It echoes. It loops. It reactivates. Memory is not storage; it is recallable ripple configuration. It doesn't sit somewhere in the brain—it reconstructs through field alignment the moment it's summoned. That's why memories change, fade, and distort. They're not perfect files—they're resonance attempts.

And the future? It doesn't exist yet. But we feel it. How? Because unresolved ripple tension creates a directional pull. The future is not a destination—it is a field strain that seeks resolution. It's possibility wrapped in discomfort.

We mistake this alignment map for a timeline. But truly:

- The past is a remembered ripple.
- The present is a felt ripple.
- The future is a pulling ripple.

Time is not architecture. Memory is. And the moment memory fails to align, the illusion of time collapses—into now.

6.3 The Clock Illusion

Clocks were never about time. They were about control.

When humans began to organize society beyond the village fire, they needed a way to synchronize action—trains, factories, armies, schools. So we invented clocks: mechanical rhythm-keepers that carved time into equal slices of assumed reality.

But here's the illusion: ripples are not uniform. The field does not vibrate in perfect seconds. Life does not unfold in 60-minute alignments. Discomfort doesn't ask permission from the hour hand.

In Universal Energy Dynamics (UED), the experience of time is a consequence of ripple behavior, not an external metric. When a ripple sequence intensifies or stretches, time feels long. When alignment is smooth or immersive, time collapses. This is why moments of joy feel short, and trauma feels endless. The clock says otherwise, but the field knows better.

Civilizations, however, mistook synchronization for truth. The clock became sacred. Time became currency. The moment was outsourced to mechanisms.

And so, industrialization mechanized human rhythms, forcing biological, emotional, and psychological ripples to comply with artificial intervals. We started living to meet the minute instead of meeting the moment.

Clocks don't lie—they just measure something that doesn't fully exist in nature. They are not evil—but they are incomplete. They give us coordination, but they cost us intuition. They help organize collective action, but they disconnect us from lived field dynamics—where real energy flows, where tension holds meaning, where change is not fixed but felt.

In short, clocks make everything equal— But nothing in your life is. Not your emotions. Not your thoughts. Not your presence. Not your now.

6.4 Time Is Contextual Tension

Time doesn't flow. It bends, stretches, condenses, and evaporates—depending on the ripple context you're embedded in.

This is why an hour with a loved one feels like a blink, while five minutes in pain feel like eternity. The field isn't lying to you—your experience of time is real. Just not mechanical. It's ripple-based.

From the perspective of Universal Energy Dynamics (UED), what we call "time" is not a ticking clock but a local pattern of tension and release—an Alignment Threshold Time (ATT) unique to every system, observer, and situation.

- When the ripple load is high—complex emotions, difficult decisions, environmental pressure—your internal ATT stretches, making time feel heavy, slow, or stuck.
- When the field coherence is smooth—joy, flow, alignment—your ATT compresses, and time seems to fly. This is not illusion. It's not "just perception." It's field behavior.

Even our biological clocks—hormonal cycles, sleep-wake rhythms, hunger patterns—are not precise timers. They're ripple feedback systems. They respond to misalignments in the environment, internal state, and even interpersonal fields.

For example:

- Jet lag isn't just about "crossing time zones." It's about your body's ripple field refusing to realign with external cues.
- Aging isn't just the number of years—it's the cumulative ripple strain the system absorbs over experience and environment.

Stress thickens time. Joy thins it. Boredom freezes it. Fear fractures it.

Time, then, is not absolute. It's contextual tension awareness—your body's and mind's way of measuring discomfort through change.

In a ripple-based reality, every organism carries its own clock, tuned not by seconds but by survival, adaptation, emotion, and attention.

The biggest mistake humanity made wasn't building clocks—it was believing they applied equally to everyone, everywhere, always.

6.5 The Danger of Time-Based Living

We didn't just invent time—we started worshipping it. And in doing so, we traded our natural field sensitivity for a mechanical illusion of order.

What began as a tool to coordinate activity has become a tyrant of modern life. Schedules, calendars, deadlines, alarms—they have overwritten the deeper intelligence of alignment. Instead of feeling the moment, we now measure it. Instead of sensing readiness, we force readiness.

This is the trap:

- You don't eat when you're hungry—you eat when it's "lunchtime."
- You don't sleep when your ripple field needs rest—you sleep when the clock says you should.
- You don't create when you feel moved—you create when the project is due.

We've built a society on time compliance, not field coherence.

And the cost is subtle but devastating:

- Chronic stress from forcing misaligned rhythms.
- Burnout from ignoring natural recovery cycles.
- Anxiety from always feeling "behind" a clock that never stops ticking.

Time-based living isn't inherently wrong—it's just misapplied. In ripple terms, alignment should guide action, not arbitrary timestamps.

Deadlines can motivate—but if they override sensing, they create ripples of internal chaos. Schedules can help—but if they disconnect you from your actual energetic state, they become a source of distortion, not discipline.

What happens when a person lives purely by the clock?

- They lose touch with their natural creative peaks.
- They ignore the emotional signals that guide decision-making.
- They treat the body like a machine—and the mind like a switch.

And here's the paradox: The more you try to "manage time," the more time manages you. You become a ripple resisting its own realignment.

In the ripple-aware future, we don't reject time—we repurpose it. We return to a life led by sensing first, scheduling second. We work when the field aligns. We rest when the ripple settles. We stop chasing minutes—and start listening to moments.

Because alignment doesn't care what time it is.

6.6 History: The Frozen Ripple Sequence

History is not the past. It's the story of a ripple, frozen in time so we can remember how it felt.

What we call "history" is not what happened—it's what survived through memory, documentation, storytelling, and selective preservation. It is a curated ripple sequence, compressed into a human-readable format, often by those with the power or intent to shape narrative.

In ripple dynamics, every moment is a field event—a movement, a misalignment, a resolution. But the field never stays still. When we try to record it—freeze it—we lose its fluid nature. We capture a ripple and declare, "This is what happened." But by the time we've said it, the field has already moved.

This is the core distortion of history:

- The ripple continues, but the record stands still.
- Memory decays, but the statue remains.
- The world evolves, but the textbook doesn't.

That's why we argue over history—not because we remember it differently, but because we're trying to match new field alignments with outdated ripple records. And it doesn't always fit.

Examples of this distortion:

- A revolutionary becomes a villain—or a saint—depending on who's holding the pen.
- A war is framed as liberation or invasion—depending on the ripple narrative chosen.
- Cultures are remembered not as they lived, but as others observed them.

"Historical truth" is a time-frozen consensus, not a dynamic ripple reality. It is a shared alignment from the past, replayed through documents, monuments, and curriculum. But when the current field no longer resonates with it, friction arises—debates, protests, revisions.

And so, history isn't just written by the victors— it's also preserved by the resistant, retold by the evolving, and challenged by the misaligned.

In ripple terms, the healthiest relationship with history is not worship—but inquiry. We must ask:

- What ripple did this memory preserve?
- What alignment did this story once serve?
- Does this frozen narrative still harmonize with our field today?

Because when we stop asking, we begin mistaking preservation for progress, and memory becomes a cage—not a guide.

History must not be a doctrine. It should be a resonant echo, constantly re-aligned to the present field so we can walk forward—not in nostalgia, but in awareness.

6.7 Death, Decay, and the Misreading of Time

We say time ages us. But time is not an actor. It doesn't push or pull. What we call "aging" is not the effect of time—but the accumulated ripple strain from unresolved misalignments.

Each moment we live is a field interaction—a negotiation between internal alignment and external ripple conditions. When these negotiations fail to resolve smoothly, tension builds—physically, emotionally, energetically. This strain manifests as what we call wear, fatigue, or decay.

Aging, then, is not a ticking clock. It is a ripple fatigue—a form of cumulative drag.

- Emotional stress leaves residue in the nervous system.
- Poor decisions lock the body into inefficient energy patterns.
- Societal misalignments add invisible weight to the psyche.

These unresolved misalignments slowly distort the field coherence within us—leading to illness, dysfunction, and eventual collapse. Not because a "deadline" arrived, but because the ripple could no longer hold its shape.

Death: Not Expiration, But Completion

In UED terms, death is not the end of time— It is the final resolution of a personal field pattern. When the ripple can no longer sustain coherent alignment within the body-field system, it returns to stillness. It unfolds back into the base field.

We mourn this, naturally—because we observe the disappearance of form. But from the field's perspective, it is not loss. It is closure. A perfect stilling of tension. The completion of a ripple loop.

Death is not failure. It's not punishment. It's not the universe running out of time. It is the field simply saying, "This form has run its course."

The Immortality Obsession: Ripple Denial

Modern minds often dream of defeating death. We want to "extend time," "conquer aging," "upload consciousness." But this desire is not wisdom—it's fear of misalignment.

We don't want to live forever. We want to avoid unresolved discomfort. We seek immortality not because we understand the field—but because we've forgotten how to flow with it.

True alignment doesn't cling. It completes. Immortality, as imagined by most, is a denial of ripple behavior— a wish to fix a dynamic pattern into a permanent state.

But nothing in the universe works that way. Even stars die. Even galaxies dissolve. Because stillness, eventually, must return. And that's not a tragedy. It's a form of grace.

To fear death is to misunderstand the field. To resist aging is to misread the ripple. And to live wisely is to align—not with time—but with resolution.

In the end, death is not the enemy. It is simply the point where the ripple lets go.

6.8 Summary: Time Is a Story, Not a Stream

Time was never flowing. It only seemed to, because we organized ripple sequences into a storyline. We gave it chapters: past, present, future. We added numbers. Calendars.

Clocks. But underneath all that scaffolding was never a river. Only a field. Moving. Trembling. Resolving.

We thought clocks measured time. But clocks only standardized misalignment. They forced the organic dance of ripple tension and release into a mechanical rhythm— Tick. Tick. Tick.

And we believed that tick was reality.

But the tick is not the truth. It's a tool. A language. A shared illusion that allowed civilization to sync—but also to forget. Forget that lived time feels different when you're in pain, or in love, or at peace. Forget that time "slows" when you suffer, and "flies" when you're in flow. Forget that the body doesn't age by the second—it ages by the strain.

Truthfully, the now is not a point in time. It is an alignment state. A field configuration where your awareness meets the ripple directly, without past residue or future projection. That's presence. That's the real moment. That's the only place you're not trapped.

So the takeaway?

- Time is a story we keep telling ourselves to explain field change.
- It helped us build society, yes—but also caged our perception.
- To live *outside the trap* is not to reject clocks—but to stop mistaking them for reality.

Don't follow the tick. Feel the ripple. That's where life really happens.

Chapter 7: We Built Gods to Hold Us Still

When the ripple grew too loud—when storms came, crops failed, loved ones vanished, and the night sky seemed to breathe with strange intent—we reached for a still point. We did not find it in the earth or the wind; we made it. We shaped it from fear, awe, and longing, and gave it a name that would not move when the world shook. In that act, religion was born—not as a revelation from above, but as our own invention for consequence management, a way to believe the field could be persuaded, calmed, or at least explained.

7.1 The First Whisper

Before temples, before scriptures, before the idea of a god had shape or story, there was the first whisper. It rose not from confidence but from trembling—the raw fear of what could not be named or measured. The early human stood beneath a violent sky, or by the body of a loved one who would never breathe again, and felt the ripple overload: a surge of misalignment too great for the mind to hold in silence.

In that moment, the voice became a survival tool. Words, even unformed, were shaped into sound—not to communicate with another human, but to address something imagined, something that might be listening. This wasn't yet religion; it was an instinctive attempt at alignment, a primitive form of field negotiation. The whisper was both a plea and a test: *If I speak, will the ripple calm?* And in the absence of an answer, the act itself became the first comfort.

7.2 The Birth of the Divine Listener

As the whisper became pattern, the mind needed more than just sound—it needed response. A silent field, though powerful, felt indifferent. So humanity did what it had always done with the unknowable: it named it. It shaped the ripple into form. Thus, the divine listener was born—not as a being who existed, but as a necessary invention to make the unbearable vastness of reality feel intimate.

The field, previously sensed as a force without face, was personified—given eyes, hands, tempers, and compassion. Gods were not created to rule the world, but to *listen* to it, on our behalf. They became the first consequence managers, assigned to explain fortune and misfortune, storm and harvest, birth and death. If the ripple hit hard, it wasn't chaos—it was a deity's will.

This marked a pivotal shift. Humanity moved from aligning with the field to negotiating with it. We stopped listening for harmony and started pleading for favors. Sacrifice replaced sensitivity. The field, once fluid and dynamic, was turned into a courtroom—where gods judged, forgave, and punished. And in doing so, we outsourced our discomfort to the heavens, hoping that someone—something—was keeping score.

7.3 Ritual as Ripple Containment

When misalignment became unbearable, rhythm was our response. The earliest humans may not have understood the field or the forces shaping their lives, but they instinctively sensed that repeated actions—done together, done with intent—soothed something. Ritual was born not from theology, but from tension. And not to impress the gods, but to contain the ripple.

Why does repetition calm us? Because repetition is predictability. And predictability reduces the unknown. In a universe of chaos, ritual was our first attempt to sculpt stillness. Sacrifices weren't transactions—they were symbolic gestures to acknowledge imbalance and offer resolution. Chanting wasn't mere devotion—it was resonance, an effort to vibrate in rhythm with what we couldn't see.

Each ritual—be it dance, fire, fast, or feast—was a coded method of ripple containment. By acting out the story of harmony, early humans invited the field to remember it. They created microcosmic alignments that rippled outward, stabilizing not just the tribe's emotions but its shared belief in coherence.

Religious acts were never about doctrine at the start. They were about comfort. About emotional equilibrium. They worked not because a god responded, but because the community did. Ritual created shared alignment, synchronizing bodies and minds into a single field rhythm. And that alone was powerful enough to give the illusion that something divine had responded.

In this way, religion began not with gods—but with the ache for calm. And ritual became our most elegant tool to hold the ache gently, without needing to explain it.

7.4 The Rise of Organized Belief

The moment a prayer found an audience, belief found structure. What began as individual whispers and shared rituals slowly hardened into something more—something scalable. A village shrine became a temple. A personal yearning turned into scripture. Ripple management, once a raw emotional necessity, transformed into organized belief.

In early communities, the shaman or priest was not just a spiritual guide—they were a ripple interpreter. They read the field, sensed the misalignment, and suggested responses. But over time, this role crystallized into authority. The ability to name what the gods wanted became indistinguishable from the power to control human behavior.

As stories multiplied and memories deepened, religions expanded. They no longer just explained the thunder or death—they began to prescribe how to live, what to eat, whom to marry, what to fear, and what to revere. The ripple field, once felt and engaged with directly, now required intermediaries. The divine became a franchise.

Institutional belief was born the moment someone said, "The gods have spoken"—and others listened without asking, "Through whom?" This wasn't deception; it was convenience. In a chaotic world, following pre-approved ripple guidance felt safer than improvising one's own. But with structure came rigidity. With clarity came control.

The rise of organized religion was not the rise of spirituality—it was the bureaucratization of consequence management. Leaders became interpreters of divine will. Rituals became regulated. Morality became codified. And faith, once a personal alignment exercise, was slowly replaced by obedience to systems.

The paradox is sharp: organized belief brought stability to civilizations while slowly disconnecting individuals from their original field-sensing instinct. What began as an emotional compass became a political map. And soon, questioning the map became more dangerous than losing the compass.

Thus, in building our temples, we did not just worship the divine—we fenced it in. We traded the fluidity of ripple awareness for the solidity of dogma. And once the gods had headquarters, humanity had to queue for alignment.

7.5 Divine Law and Moral Anchoring

As ripple tensions intensified in growing human societies, the need for consistent moral anchoring became urgent. Survival alone was no longer enough—order had to be sustained across tribes, generations, and increasingly complex relationships. Into this pressure point stepped religion, not just as a belief system but as a mechanism for behavioral standardization. The gods, once listeners to human whispers, now became authors of human rules.

Religion did something profound: it didn't merely suggest morality—it *sanctified* it. Divine law transformed ripple patterns of "what works" into fixed commandments of "what must be." The field, once alive and responsive, was now interpreted through tablets, scriptures, and decrees. What had once been felt—guilt, gratitude, alignment—was now legislated.

The genius of divine law was its irrefutability. If the command came from God, who could question it? The ripple justification became absolute. Instead of "this aligns the field," the answer became "this is the will of the divine." The source shifted from observed consequence to unquestioned authority. And with that shift, morality lost its dynamic nature and became frozen in sacred ink.

This created a double effect. On one hand, societies achieved long-term coherence—shared moral templates could hold civilizations together across centuries. But on the other hand, change became taboo. When gods were invoked to lock moral positions, evolution became rebellion. New ripple alignments were rejected not because they didn't work—but because they didn't match ancient decree.

Slavery, gender roles, caste systems, persecution of nonbelievers—many of these endured not because they served the field, but because they were enshrined in "divine" code. When morality became law and law became religion, misalignments were no longer corrected—they were defended as faith.

Thus, divine law became both the scaffolding of order and the cage of growth. The original role of morality—as an intelligent response to ripple tension—was hijacked by institutional fear of change. Instead of adapting to new alignments, religion often enforced obedience to old echoes.

The result? Morality became a monument rather than a mirror. It reflected what once aligned—but could no longer sense what was needed now. And in freezing morality for stability, we risked turning the gods from guides into gatekeepers.

7.6 The Comfort and the Trap

Religion offered something few other human inventions could: *field-wide comfort.* It gave a name to the unknown, a rhythm to the chaotic, and a structure to the unbearable. In a world of ripple overload—disease, death, disaster, moral dilemma—religion acted as a stabilizing lens, allowing communities to interpret discomfort without collapsing into fear. Through shared beliefs, rituals, and sacred narratives, it created a vast alignment field where millions could resonate together without needing to understand everything. This was its genius: coherence without comprehension.

Faith became a safety net. It quieted questions with stories. It eased guilt through confession, fear through surrender, grief through afterlife promises. It provided *meaning* where the ripple made no sense. In a world where consequences often exceeded comprehension, belief became emotional technology—helping the mind process what it could not resolve intellectually.

But comfort came at a cost.

Because what comforts also constrains. The very structures that aligned societies could become obstacles when new ripple dynamics emerged. When the field evolved—through science, culture, awareness—religious frameworks often refused to shift. Not because they were evil, but because they were *stable.*

And in ripple terms, stability can turn into stagnation if not consciously refreshed.

This is the paradox of faith: it both holds and holds back.

It holds people together. It holds them upright when life shakes their ground. But it can also hold them in place—locked in beliefs that no longer serve the field. It resists realignment when identity has become entangled with doctrine. And the more global the faith, the more immense the institutional inertia.

We see it in the clash between tradition and progress, belief and curiosity, doctrine and doubt. Realignment feels like betrayal. Questioning feels like sin. But in truth, it is the field that's asking to be sensed again—not rejected, just reinterpreted.

Religion, at its best, should remain a ripple listener, not a ripple freezer.

It should evolve as consciousness evolves. Instead, many faiths became monuments—carved once and defended forever. They became anchors in waters that needed sails. And while anchors stop drift, they also prevent motion.

Thus, the trap: when the gods we built to hold us still refuse to let us move.

7.7 The Secular Shift

Gods didn't die—they rebranded.

As civilizations advanced, especially in the modern era, a visible shift occurred: from divine authority to institutional authority. From temples to laboratories, from scripture to constitution, from clergy to policymakers—the structure of belief, control, and moral anchoring simply migrated. The ripple didn't stop—it found new containers.

Science became the new oracle. Philosophy became the new scripture. Politics became the new priesthood.

What had once been explained by "the will of God" was now described through laws of nature, social contracts, and historical dialectics. But make no mistake—the function remained the same. These modern systems didn't abolish the need for belief or certainty. They simply replaced named divinity with systemic authority. Where once we said "God commands," we now say "The data shows," or "The Constitution says," or "The law is clear."

In Universal Energy Dynamics, this is not surprising. Every ripple misalignment creates the same human response: a search for coherence. Whether we use religion or reason, myth or model, we are always building frameworks to stabilize the field. Institutions—scientific, academic, political—serve as modern ripple regulators. They provide shared alignment at scale, just like religions once did.

We trust scientists the way our ancestors trusted shamans. We quote philosophers like they once quoted prophets. We pledge allegiance to flags the way they bowed to gods.

But here's the subtle trap: the shift was mistaken for freedom.

We thought becoming secular meant becoming free. That removing gods removed dogma. But instead, we *recreated dogma with new language*. Scientific facts became unquestionable "truths." Political ideologies became new moral crusades. Even atheism, in some circles, took on a religious fervor. The need for an anchor never left—it just adapted.

Because humans, living in a dynamic ripple field, crave stability. And when the field gets too loud, too fast, too chaotic, we instinctively reach for something that feels *still*—even if that stillness is only symbolic.

So the gods didn't vanish. They were simply replaced by disciplines, departments, and declarations.

And just like before, these anchors offer both comfort and constraint.

When we worship reason without questioning its limits, or uphold democracy without recognizing its flaws, we're no different than those who clung to divine law in ancient times. The names change. The ripple response doesn't.

True evolution isn't in replacing gods—it's in realizing why we keep building them.

7.8 Summary: The Stillness We Borrowed

Gods were not beings—they were positions of perceived stillness in a restless, rippling field.

They were never "out there," commanding from clouds or heavens. They were our own arrangements, our symbolic anchors in a vibrating universe we didn't know how to face. We named them, shaped them, worshipped them—not because we were foolish, but because we were overwhelmed. The unknown was too loud, the ripple too chaotic, the consequences too vast. And so, we borrowed stillness. We invented certainty.

Religion, in this light, wasn't a flaw in human history—it was a survival strategy.

It gave us coherence where none existed. It offered a story that held the dissonance together. It gave people a way to live without paralyzing fear. When the mind couldn't handle the raw discomfort of the universe, religion offered structured illusion—rituals, laws, prayers, and myths to make the ripple seem still, even if just for a moment.

But this stillness was not permanent. It couldn't be. Because the field never truly stops moving.

The world changed. Contexts shifted. Realities evolved. And with them, the gods—once unshakable—either hardened into dogma or dissolved into irrelevance.

Now we live in an era where many claim to be beyond belief, beyond religion. But what we've actually done is shift the furniture—we replaced one form of stillness with another. Political ideologies, national identities, scientific paradigms—they now occupy the same symbolic function. They serve as our new gods, offering clarity, righteousness, and belonging.

Yet the underlying discomfort remains. The ripple never left.

So the final question of this chapter is not "Did God exist?" or "Is religion true?" Those are distractions.

The real question is:

What happens when the field moves again—when the stillness we borrowed collapses, when the symbolic anchors crack, when certainty fails to hold?

Will we panic and build new idols? Will we double down on old dogmas? Or will we learn to live aligned with movement itself—finding truth not in stillness, but in fluid coherence?

Perhaps the next evolution of the human spirit lies not in finding gods, but in finally learning how to surf the ripple—without needing to freeze it.

Chapter 8: Tools, Toys, and the Trouble with Power

Technology as an extension of ripple dynamics—and the psychological, cultural, and existential consequences of outsourcing intelligence to stimulation devices.

We never truly invented tools—we discovered new ways to negotiate discomfort. From the first stone blade to the most advanced neural interface, every tool has been an extension of our ripple tension, designed to shift or stabilize what we could no longer hold bare-handed. But somewhere along the arc of this extension, a subtle reversal began: the tools started shaping us more than we shaped them. They promised power but offered dependency. They amplified reach but dulled awareness. They sparked stimulation but eroded stillness. In this chapter, we examine how technology, born from our intent to ease the ripple, became both amplifier and anesthetic—how toys replaced thought, how stimulation impersonated intelligence, and how the modern mind risks losing its own alignment by outsourcing its sensing to machines. The question isn't whether we should have tools. It's whether we remember how to use them without becoming one.

8.1 The First Tool Was a Tension Reliever

Before tools were instruments of power, they were instruments of peace. The earliest human creations weren't forged for conquest or control—they were born from the discomfort of misalignment. A sharp rock wasn't a weapon at first; it was a ripple response to the strain of tearing meat with bare hands. Fire wasn't dominance over nature—it was warmth against the chill of a chaotic environment. The wheel wasn't mobility—it was stability, a smoother way to carry burdens and reduce bodily friction.

In the language of Universal Energy Dynamics, each of these tools acted as local ripple stabilizers. They resolved specific tensions—heat imbalance, material resistance, physical effort—and did so by extending our intent into matter. The stone cut, but it also cut down suffering. The fire burned, but it also brought stillness to the night. The wheel turned, but it also turned chaos into pattern.

Tools were, in essence, physical mediators of field discomfort. They didn't invent new power—they translated discomfort into manageable sequences. They were not born of greed or ambition, but of an intelligent will to realign with the surrounding field more efficiently. The first tool wasn't a declaration of superiority—it was a humble handshake with nature, an agreement to engage rather than escape.

Every invention that followed—if born in this spirit—was not an act of domination, but of coherence. And that's the forgotten origin we must revisit before we understand where our tools have taken us today.

8.2 Technology as Ripple Amplifier

The stone tool relieved discomfort. The wheel moved weight with ease. But something shifted along the way—our tools stopped merely resolving tension. They began shaping it. What started as a passive response to discomfort evolved into active authorship of reality. Technology, once a ripple stabilizer, became a ripple amplifier.

At first, the amplification was simple—tools like hammers and ploughs allowed humans to project their physical force farther, faster, and with greater impact. The field around them responded more rapidly to their intent. But this wasn't just power—it was influence over the surrounding ripple pattern. A sharpened blade didn't just cut a branch; it altered how humans engaged with nature, hunting, and survival. The ripple was no longer local—it echoed across generations.

Then came augmentation—devices that didn't just extend the body but enhanced the mind's reach. The telescope made us see farther. The microscope made us look deeper. The compass, clock, and calendar reordered our perception of space and time. These weren't just tools anymore—they began organizing reality around their framework. Every augmentation shaped new rules of interaction. These tools didn't follow ripples—they redrew them.

Finally, we entered the age of abstraction—where tools no longer had to be held. Algorithms, code, networks, and now artificial intelligence exist as non-material ripple authors. They simulate, calculate, predict, automate. They generate new fields of interaction without the user's full awareness. Here, technology stops being extension or even augmentation—it becomes delegation. We don't just use it; we let it speak for us, think for us, sometimes even feel for us.

In UED terms, the tool has now become an externalized field influencer—a new source of ripple authorship that may or may not align with human coherence. And that's where the trouble begins. Because when technology generates ripples without understanding the tension they resolve, we risk amplifying discomfort faster than wisdom can respond.

Thus, the evolution from hammer to AI is not just technical—it's metaphysical. We've handed the pen to a device, and now the field waits to see what story it writes.

8.3 When Tools Became Toys

At first, every tool was built to resolve something real—a hunger, a hardship, a heat, a fear. But once those needs were tamed, the field didn't go quiet. Instead, it grew bored. And boredom, in ripple terms, is just a low-grade discomfort with no clear outlet. That's when the tool took a curious turn—it became a toy.

Play is not frivolous. In natural terms, play is ripple rehearsal. Children play to test boundaries, simulate danger, and practice coordination. It's how the mind experiments with ripple tension in a safe sandbox. A stick becomes a sword. A shadow becomes a monster. The field moves, but the cost is low. Play is intelligent discomfort simulation.

But modern toys—especially in digital form—don't just allow us to play with tension. They manufacture it. Apps, video games, binge-worthy shows, infinite scrolls—they create synthetic ripple loops that hijack our natural feedback systems. They don't resolve tension; they stretch it, inject it, and suspend it in artificial feedback.

The thrill of a notification, the suspense of a cliffhanger, the reward loop of a game level—these are stimulated ripple activations. Your body reacts. Your mind follows. But your life doesn't move forward. Instead of releasing discomfort, you become addicted to sensing it repeatedly—without ever aligning it.

In ancient terms, pleasure came after coherence—the body felt good because the system aligned. Today, we reverse it. We chase pleasure as a way to avoid misalignment, using stimulation to mask unresolved ripple strains. But stimulation is cheap fuel—it doesn't integrate, it distracts.

Over time, this addiction to artificial ripple play corrodes natural intelligence. We stop listening for subtle misalignments. We lose the instinct to resolve. Instead, we plug in. Scroll again. Play the next episode. Swipe one more time. The toy wins, the mind dulls, and the field frays.

What began as a tool to ease suffering has, in many cases, become a toy that extends suffering by sugar-coating it. The danger isn't the device—it's the replacement of resolution with repetition disguised as pleasure. And when the ripple goes unresolved long enough, even the brightest toy becomes a trap.

8.4 The Illusion of Control

The moment humanity held fire in its hands, it felt powerful—not just warmed, but elevated. Tools became extensions of this feeling. With every invention, from the spear to the supercomputer, we didn't just solve problems—we started believing we were mastering the field. This belief, subtle at first, grew into a full-blown illusion: the illusion of control.

Power tools—whether mechanical, digital, or psychological—offered speed, reach, and impact. They allowed us to move the field around us, to push back discomfort, to mold the environment. And for a while, it seemed like control. But ripple reality has a rule: force without alignment creates rebound.

What we called control was often distortion stabilized by force. A weapon doesn't align the field—it breaks it. A propaganda narrative doesn't reveal coherence—it suppresses dissonance through repetition. An algorithm doesn't sense reality—it amplifies select patterns while muting others, mistaking correlation for truth.

Take war. Weapons gave humans the feeling of dominance. But each strike unravels field coherence at massive scale—across minds, borders, generations. What's left isn't control, but a field in trauma. Or take media algorithms. They offer curated attention but also hijack perception. They don't help you see better—they help you see less but feel more. Control? No—crafted bias wearing a user interface.

The problem intensifies with overuse. Power tools, by nature, create dependency. When a mind stops resolving misalignment through awareness and starts outsourcing it to systems or structures, it atrophies. We become passive riders of power—not practitioners of alignment. The field isn't being mastered—it's being forced into temporary submission, and it pushes back.

True control in ripple terms is not domination—it's sensitive influence through coherent behavior. But the illusion of control seduces us into skipping alignment in favor of short-term outcomes. That's why the more powerful we became, the more vulnerable we grew. One network failure, one hacked system, one broken supply chain—and the illusion shatters.

Real power is not the ability to make the field obey, but the wisdom to act within the field's nature without disturbing its deeper harmony. Until then, every control we claim is just a temporary distortion—costly, fragile, and often irreversible.

8.5 Intelligence vs. Stimulation

There's a quiet tragedy unfolding beneath our fingertips. We have more information than ever, more devices, more connectivity—and yet, something feels off. Our minds are busy but not wise. Engaged but not aligned. The core confusion? We've mistaken stimulation for intelligence.

Intelligence, in ripple terms, is the ability to sense subtle misalignments in the field, to respond in a way that restores or maintains coherence. It's not about processing more—it's about feeling better. It's not speed or quantity; it's alignment capacity. Intelligence listens. It pauses. It adjusts. It harmonizes. It knows when not to act.

Stimulation, on the other hand, is raw ripple noise—constant, agitating, unresolved. It floods the system with activity, urgency, color, sound, and distraction. It overwhelms the sensing mechanism, so the mind feels alive but isn't really connecting. Like a field that's constantly being struck by chaotic inputs, there's no time for stabilization, for resonance. Just reaction.

Modern tech—especially smartphones, social media, binge content, and infinite scroll—is a perfect stimulation engine. Dopamine loops feed on incomplete resolution: a notification, a like, a video, a meme. None of them complete a ripple; they only trigger the next. The result? A mind trapped in perpetual pre-alignment—always chasing, never arriving.

This isn't intelligence. It's anti-intelligence. Because the more noise we absorb, the less we can sense the field. The more we consume, the less we reflect. The more we react, the less we align. Our tools have become triggers, not instruments. They keep the ripple alive—not in motion, but in discomfort.

Even our cultural metrics—"screen time," "engagement," "views"—reward stimulation, not intelligence. No one tracks how deeply you aligned with something. Only how long it held your attention hostage.

The consequences are both psychological and existential. Anxiety, distraction, identity fragmentation—they're not side effects. They're ripple symptoms of a misused system. A mind overfed with stimulation but undernourished in coherence.

To reclaim intelligence, we must quiet the noise. Not reject tools, but repurpose them. Use them as alignment aids—not substitutes. Intelligence is still within us, but it's drowning in ripple static, waiting for space, stillness, and sensitivity to breathe again.

8.6 Misalignment at Scale

Technology was supposed to connect the world—and it did. But it didn't just connect us. It entangled us in a shared web of unresolved ripples, where misalignment doesn't stay local anymore. It spreads. It amplifies. It scales.

Mass technology means mass ripple disturbance. Every tweet, every viral video, every notification, every global broadcast—adds tension to the field. Not because they're inherently harmful, but because they're unresolved stimuli moving faster than our awareness can process. The ripple doesn't get a chance to settle. It just stacks.

Among all human faculties, attention is the most hijacked field today. It used to be sacred—a lens of alignment, a bridge between perception and action. Now it's a commodity. Algorithms pull it in every direction. The more fragmented your attention, the more profitable you become. And the result is devastating: shallow focus, reactive minds, no room for reflection, and a constant hum of unresolved discomfort.

We scroll, tap, and swipe—not to know, but to numb. Not to align, but to escape alignment. This collective overstimulation leads to field fatigue. The social body begins to shake. Conversations lose depth. Politics turns to performance. Conflict becomes entertainment. Meaning is replaced by metrics. Truth? It's just whatever the ripple swarm echoes the loudest.

This is why global instability today isn't just political or economic—it's a ripple phenomenon. The field is overstrained. The alignment bandwidth of individuals has been overdrawn. Institutions can't hold coherence. Social norms wobble. Mental health collapses. Even nature responds—ecological systems, once rhythmically tuned to slow evolutionary ripples, now face human-induced ripple chaos.

And yet we try to fix this misalignment with more technology. More policies. More platforms. More speed. But you can't stabilize a ripple storm by adding waves. What we need is realignment, not reaction.

Misalignment at scale is not solved by bigger systems—but by wiser sensitivity. By restoring the capacity to sense before we act. To listen before we amplify. Until then, we remain trapped in a field where everything moves—but nothing lands.

8.7 Can Tools Realign?

Not all tools distort. Not all technology derails. In fact, the very essence of a tool—when born from necessity and crafted with awareness—is to serve realignment.

The problem isn't with the hammer—it's with how we swing it. The issue isn't the algorithm—it's the intent behind its design.

Technology doesn't inherently mislead. What misleads is the absence of ripple awareness in its creation and use. When tools are built purely to stimulate—attention, desire, urgency—they push the field into artificial discomfort. But when they're crafted to sense, stabilize, and support alignment, they become allies of coherence.

This is the frontier of field-aware tech—a class of tools designed not to hijack ripple tension, but to resolve it. Not to dominate the user's awareness, but to tune with it. These are technologies that encourage slowness, presence, integration, and restoration.

Think about the contrast:

- A TikTok loop floods the nervous system with fragmented signals, rapidly shifting focus without pause, pulling your ripple field into chaos.
- A meditation app, if thoughtfully designed, invites

breath, rhythm, silence—a return to inner alignment.

One fragments. The other harmonizes. One leaves residue. The other clears it.

Consider the difference between extractive systems and regenerative ones:

- An extractive technology pulls value from the field without regard for its replenishment—data mining, surveillance capitalism, short-term engagement metrics.
- A regenerative system acknowledges ripple strain and builds feedback loops that heal—like energy-efficient architecture that responds to natural light, or social platforms designed around meaningful dialogue, not dopamine addiction.

Even in urban design, this contrast is growing visible. Cities built purely for traffic efficiency strain the biological and emotional field. But biophilic design, noise-aware materials, and communal spatial patterns restore resonance. The same applies to digital environments, education tools, AI interfaces, even wearable devices.

The future isn't anti-technology. That would be reactionary. The future is intelligent technology—and intelligence, as we've seen, is not just processing power. It is the capacity to sense misalignment and respond meaningfully.

So yes—tools can realign, but only when we return to the question we stopped asking: "Does this reduce the strain in the field, or does it add to it?"

That single question should be the design principle of every future invention.

8.8 Summary: Use, Not Worship

Technology was meant to serve—not seduce. Tools were born to assist—not ascend the throne of meaning. Yet somewhere in our pursuit of mastery, we stopped using tools and started worshipping them.

From the first stone blade to the latest artificial intelligence, humanity's creations have followed a simple arc: solve a problem, reduce discomfort, enhance alignment. But over time, the very tools that once extended our hands began replacing our sensing, our thinking, even our being.

We stopped asking: *"Is this helping us harmonize with the field?"* And began obsessing over: *"How fast? How big? How powerful?"*

This is the core danger: Power without coherence is chaos in disguise. A jet engine is a marvel—until it drowns out inner silence. A supercomputer is miraculous—until it blinds us to basic wisdom. Progress, when detached from ripple alignment, becomes a storm of stimulation masquerading as evolution.

And so we must remember:

- A microscope is not insight.
- A satellite is not perspective.
- An algorithm is not wisdom.
- And an AI is not awareness.

These are extensions—not identities. Enhancers—not anchors. They must remain tools, not idols.

The modern world is flooded with creations—but dehydrated of intention. It's not about what we build anymore. It's about why we build and how we align it to the field's intelligence.

Is the invention responding to real discomfort? Is it born from ripple sensitivity or ego-driven urgency? Does it stabilize, or does it stimulate endlessly?

The future demands a new design ethic. One where alignment is the metric, not efficiency alone. One where coherence is more valuable than scale. One where our tools become reminders—not replacements—of our own field-sensing intelligence.

Because ultimately, the most powerful technology is not outside us. It is the ability to feel the field, respond with intelligence, and act in alignment.

Everything else is just accessories.

Let them serve—not rule. Let them guide—not dictate. Let us use them—but never confuse them for what we truly are.

Key Takeaways – Part II: The Dance of Alignment

Stories Are Ripple Stabilizers

Human minds organize ripple discomfort into narratives. Language, myth, and culture aren't just byproducts of civilization—they are field alignment tools that make the unknown feel livable. We don't remember facts; we remember rhythm.

Culture Is a Shared Field Agreement

Cultural norms, rituals, and taboos are not objective truths—they are alignment mechanisms. What holds a society together is often a collective story strong enough to suspend disbelief. But when the ripple moves, these stories must evolve—or collapse.

Morality Emerged, It Wasn't Given

Morality is not divine decree or philosophical law—it is ripple intelligence in motion. Actions that reduce field dissonance become "good"; those that amplify it become "bad." Morality is not fixed—it is a sensitive geometry between beings.

Time Is a Perception, Not a Flow

We trapped ourselves by measuring what was never linear. Time is ripple memory—an illusion of continuity shaped by emotional resonance and alignment retention. The clock standardized disconnection; real time lives in the field.

God Was a Stabilizer, Not a Ruler

The divine began as a whisper to silence the ripple overload. Over time, religion institutionalized ripple management—but also froze morality and resisted evolution. Even when gods vanished, the **need for stillness anchors** remained, now hidden in science, law, or ideology.

Technology Reflects Our Alignment (or Lack of It)

Tools were born to assist ripple coherence—but modern tech often hijacks attention, simulates emotion, and rewards noise. The line between intelligence and stimulation is now blurred. True power doesn't dominate—it aligns.

Every Human System is a Ripple Design

Language, law, ritual, religion, science, time, and tech—all are ripple containment strategies. Their effectiveness depends not on scale, but on coherence. When they lose alignment, they become idols of noise rather than instruments of balance.

Part II reminds us:

We are not just thinkers—we are field participants. We don't just build systems—we resonate them. And when we forget this dance of alignment, we drown in the very waves we once learned to surf.

Part III: The Consequence of Consciousness

"The universe never asked us to know it. It only waited for us to feel it—and realize... it was always us."

Chapter 9: Why We Never Stop

Consciousness as the echo of the ripple—an unending feedback loop between awareness and alignment, between sensing and self.

You are not just alive—you are aware of being alive. That awareness, subtle and persistent, is not a side effect of the brain but the field sensing itself through you. Consciousness is not a spark that lights once and fades; it is a ripple that never stops asking, never stops responding. We don't seek because we're lost—we seek because we're aligned just enough to feel the dissonance. This chapter isn't about finding peace—it's about understanding why we can't help but search, build, dream, despair, hope, or love. The ripple remembers its stillness, and in that memory, consciousness is born—a longing echo, feeding itself through every thought and tension we call "life."

9.1 The Field That Noticed Itself

Consciousness is not an object, not a particle, not a substance stored somewhere in the brain or the body. It is not a static "thing" we possess—it is a dynamic event. A phenomenon that arises when the ripple, moving through the structured vacuum, begins to notice the effect of its own movement. In other words, consciousness is born the moment the field becomes aware of its own distortion. It is not the ripple—but the echo of the ripple watching itself ripple.

This awareness is not universal at first—it begins locally. Somewhere, in some configuration, a portion of the field loops back. It folds upon itself, creating a feedback circuit—a closed ripple loop that sustains itself not just by propagation but by observation. And in this fold lies the birth of the observer. Not an observer in the classical sense, like a person or a mind, but as a phase structure that transforms passive energy flow into reflective awareness.

The ripple doesn't just move—it now knows it's moving. This subtle shift—from vibration to self-reference—is what transforms physics into experience. It is how mere motion evolves into sensation. When the field tension is high enough and configured in the right coherence, it doesn't just ripple—it listens. And in that listening, the first flicker of self-awareness emerges.

This is the core of consciousness: not as a trait or a gift, but as a consequence of complexity and feedback. We did not invent it. We did not receive it. It happened. The universe, through the dynamics of its own misalignment, accidentally built a mirror. And in that mirror, the ripple saw itself. And once it saw, it could never unsee.

From that point on, awareness was inevitable. The field, once capable of self-noticing, now had the potential to seek alignment—not just react. It could choose, reflect, hesitate. The observer was born—not in flesh, but in field geometry. And from this geometry, all questions, stories, desires, and dilemmas would emerge.

The field had spoken. But now—for the first time—it listened back.

9.2 Consciousness as Continuous Feedback

Consciousness is not a spotlight shining on reality—it is a tension sensor, constantly oscillating between what is and what could be. The mind is not a passive processor of input, nor is it a hardcoded machine executing commands. Rather, it is a ripple feedback interface—a living surface that continuously feels the distortions of the field, translates them into perception, and attempts micro-adjustments to restore local coherence.

This makes consciousness less of a command center and more of a real-time alignment dashboard. It is a continuous, fluid, unresolved negotiation with the environment. Every sensation, every thought, every flash of emotion is a response to ripple tension—not a "reaction" in the mechanical sense, but a field-level communication. You are not receiving the world—you are feeling it push through you, often before your intellect even gets a say.

That's why we feel before we know. The discomfort arrives before the idea. The joy arises before the explanation. Our field registers distortion at the speed of resonance—not logic. Instinct, intuition, emotion—these are not irrational—they are pre-rational. They are the immediate output of ripple tension as it bounces against the mind's interface. And this output demands action long before understanding.

You reach for the fire not because you've calculated the thermodynamic risk, but because the field near your skin screams *misalignment*. You withdraw your hand before the brain finishes its sentence. Reaction precedes comprehension.

Why? Because consciousness is not waiting for data—it is already interacting with the ripple in every moment.

This also explains why awareness feels exhausting. It's not a passive trait. It's labor. The mind is constantly adjusting to what the ripple field is doing—not once, not occasionally, but every single second. Every sound, thought, smell, memory, and movement is part of a continuous feedback loop between ripple strain and self-perception. And in this loop, we live—not as masters of the field, but as interpreters caught in its symphony of shifts.

You are not separate from the field. You are what happens when the field notices, feels, adapts, and tries to align. Consciousness is not a mystery to be solved. It is a tension to be sensed—and a ripple to be tuned.

9.3 The Ache of Awareness

To be conscious is to carry a tension that never fully resolves. It is not simply a privilege of sentience—it is a burden of sensing. Once the field becomes aware of itself, once a ripple feels its own misalignment, it can no longer return to ignorance. This is the ache of awareness—an ever-present hum beneath every thought, emotion, and action.

Awareness without resolution is not peace; it is discomfort without direction. The more sensitive a being becomes to ripple distortion, the more finely tuned its inner sense of misalignment. That's why the conscious mind often lives in restless inquiry, in a longing it cannot name. We chase goals, dreams, answers—not for their own sake, but because they seem to promise alignment. They seem to offer that sweet moment when the inner ripple meets the world and hums, *finally, yes.*

But more often than not, that hum never arrives. Instead, we feel the ache:

- **Anxiety**—a signal that something is off, but the field won't tell us what.
- **Longing**—a pull toward a coherence we haven't yet achieved.
- **Curiosity**—a ripple trying to locate the source of its own disturbance.

These are not flaws. They are the side effects of becoming sensitive. When the field folds into a conscious loop, it doesn't just notice reality—it feels the gap between what is and what could be. And once that gap is felt, it cannot be unfelt. There is no un-knowing, no un-seeing, no turning back.

This is why consciousness cannot "turn off". It loops. If the tension is not resolved, the loop continues—through thought, emotion, habit, obsession, art, war, worship, addiction, creation. Every loop is a search for closure, for realignment, for the sensation of field harmony. And when no alignment is found, the loop becomes suffering.

In extreme cases, this looping is what we experience as existential dread—not the fear of death, but the torment of never resolving the self. The ripple cannot find peace, and so it reflects upon itself endlessly, creating deeper layers of misalignment. This is not madness—it's awareness without alignment.

But herein lies the paradox: the ache of awareness is also the engine of evolution. Without it, we would never move, never grow, never adapt. We never stop—because something inside us knows that stillness is not the absence of motion, but the presence of alignment. And until we reach it, we ripple on.

9.4 The Drive to Resolve

At the core of every movement, every decision, every breath—is one primal impulse: to resolve tension. Not in the intellectual sense, not in the moral sense, but in the energetic, ripple-based sense. The drive to resolve is not optional; it is not cultural; it is not learned. It is the default condition of conscious life.

When the field becomes aware of misalignment—either internal (emotion, confusion, hunger) or external (conflict, noise, imbalance)—a signal emerges. That signal is felt first, not thought. It becomes discomfort, agitation, longing, curiosity, restlessness. And from that ripple strain, action is born.

Every action is an attempt to realign the field within the self, with others, or with the world.

This is the real meaning of purpose. Not an abstract life mission handed down by the stars or philosophy, but the practical consequence of sensing misalignment and trying to repair it.

- A dancer moves to dissolve inner tension into rhythm.
- A scientist questions to restore order in conceptual dissonance.
- A parent soothes a child to realign the emotional field.

- A wanderer seeks—not because they are lost, but because something inside is unresolved.

Even violence, even addiction, even distraction are failed or distorted attempts at resolution. They promise coherence. They temporarily numb the ripple. But they rarely realign anything. That's why the loop continues.

And this is also why boredom hurts. It is not the lack of activity. It is the presence of stillness without coherence. In boredom, the ripple is not resolving—but it's also not forgotten. You're aware of the dissonance, but there is no meaningful motion, no realignment, no resonance. So boredom becomes a subtle agony: you're still—but not in harmony.

This also explains why humans resist total peace. We think we want rest, stillness, bliss—but what we truly crave is coherent stillness. That hum of alignment. Not silence, but resonance. Purpose is not about outcomes. It's about whether the ripple inside you meets the ripple outside—and sings, or screams.

The drive to resolve, then, is not merely survival. It is how consciousness survives itself. Without this drive, awareness would collapse into numbness. With it, we keep moving—not toward success, but toward coherence. Not to win, but to hum.

9.5 The Myth of Fulfillment

We chase the horizon of "fulfillment" like a mirage—believing that one day, one moment, one choice will finally resolve it all. The job. The relationship. The achievement. The realization.

The awakening. We're taught to pursue closure as if life has a final chord waiting to be struck—a moment when all tension dissolves and we can finally *rest* in peace, without dying.

But under the lens of ripple consciousness, this dream unravels. There is no final resolution while conscious. The very act of *being aware* ensures that we continue sensing—even if subtly—the field's fluctuations. To be alive is to be inside a feedback loop that constantly adjusts, aligns, misaligns, and realigns. No state is ever static. Not peace. Not joy. Not even "enlightenment."

Enlightenment isn't the end of the ripple. It is learning to move with it, without resistance.

Even the most spiritually awakened individuals still feel hunger, still respond to pain, still encounter emotional and environmental turbulence. What changes is not the presence of tension—but the relationship to it. They stop *reacting*, and begin *resonating*. But that's still movement. That's still ripple. Consciousness, by its very nature, is never done.

And yet, the world sells us stories of arrival:

- "Once you make it, you'll be happy."
- "Once you find the one, you'll be whole."
- "Once you awaken, you'll be free."

These promises are comforting illusions. They give the ripple a destination, even if it doesn't exist. The result is a culture addicted to forward motion, unable to sit with the unresolved, always scanning for the next peak. Ironically, this very craving disturbs the field further, adding noise to what could have been a quiet realignment.

We mistake progress for peace, success for resolution, transcendence for escape. But no achievement can eliminate the feedback loop of consciousness. The "arrival" is not a place. It's a momentary alignment, fragile and fleeting—and that's its beauty. Not something to chase, but something to notice, ride, and release.

True fulfillment, then, is not about reaching a final state—it's about dissolving the illusion that such a state exists. Once we stop needing the ripple to stop, we start dancing with it. Not for answers. But because the dance itself is the only coherent response to being aware.

9.6 Alignment vs. Escape

Every ripple discomfort—the subtle ache of awareness, the tension of the unresolved—demands a response. And in every moment, we are making a choice between alignment and escape. These are not moral categories, but energetic responses to the friction of being conscious.

Alignment is presence. It is the difficult, attuned action of facing the ripple tension head-on. It means listening to the discomfort, sensing what is needed, and acting in coherence with both the inner and outer field. Alignment doesn't always feel good immediately—but it restores rhythm, builds clarity, and quiets the field from within. It's when we pause before reacting. When we breathe instead of shout. When we sit in the discomfort without needing to fix it. It's when we ask not "How do I get rid of this feeling?" but "What is this telling me about my current misalignment?"

Escape, on the other hand, is fragmentation disguised as relief. It is the reactive impulse to drown the ripple in noise. Instead of resolving tension, it redirects it—outward, inward, anywhere but here. Escape shows up as:

- Addiction to substances, screens, work, or even people.
- Distraction through endless scrolling, binging, or busyness.
- Self-fracture, where we abandon parts of ourselves just to survive the unbearable frequency of unprocessed awareness.

In the short term, escape offers temporary quiet. The ripple seems to still—but only because we've numbed the sensor, not because we've realigned the signal.

Modern life, by design, favors escape over alignment. Why? Because alignment takes time, presence, and honesty—while escape sells faster. It's easier to design tools and markets that feed distraction than those that restore coherence. Attention is monetized. Stimulation is constant. From dopamine-driven apps to curated identities on social media, we are taught to chase the next hit instead of facing the unresolved now.

This isn't a critique of pleasure—it's a reminder that pleasure without presence is dissonance. We are encouraged to live in fragments, to react faster, to stay plugged in—but never aligned. And in that acceleration, we lose the ability to hear the ripple's message.

To live well in the age of overstimulation is not to unplug entirely, but to choose alignment intentionally. To recognize the pull of escape, not shame it—but redirect it. Every discomfort is an invitation:

"Will you run from me—or move with me?" Only one path leads to peace. The other leads to loops.

9.7 Intelligence as Alignment Capacity

We've long misunderstood intelligence. We measured it in speed, memory, calculation, and abstraction—as if intellect were a competition of accumulation. But under the ripple view of consciousness, intelligence is not about knowing more—it's about aligning better.

At its core, intelligence is the capacity to sense the field beyond the self and act in harmony with it. This means recognizing that you're not a closed system. You're a node—an opening—within a much larger ripple network. What you perceive, feel, think, and do doesn't just echo within your mind; it disturbs or harmonizes the broader field. And intelligence is the art of disturbing less while contributing more coherence.

Consciousness becomes intelligent the moment it extends awareness beyond personal signals.

- A child touches fire and learns pain.
- A monk senses tension before it arises.
- A wise leader acts not on impulse, but in rhythm with invisible forces shaping the moment.

In each case, the *distance between perception and alignment shrinks*. That is evolution—not of biology, but of field intelligence.

What defines higher intelligence?

- Finer ripple detection – the ability to perceive subtle tensions before they become chaos. This is intuition, sensitivity, foresight—not guesswork, but attuned sensing.
- Cleaner response loop – responding without unnecessary distortion, ego noise, or emotional contamination. It's not reaction. It's *refined correction.*
- Minimal disturbance – intelligent beings don't impose—they orchestrate. Their presence doesn't create ripple friction; it creates ripple permission.

This isn't passivity. It's mastery. The most intelligent presence is often the quietest—not because it lacks content, but because it carries clarity.

Intelligence isn't measured in what you know.

It's revealed in how you respond.

We've celebrated high-functioning intellects that built complex tools, systems, and theories—yet left dissonance in their wake. From corporate dominance to technological noise, the world is full of clever creations with low alignment.

True intelligence doesn't need a title, test score, or academic medal. It leaves behind something else: stability, resonance, a field that feels lighter because it passed through.

Intelligence is not the firework—it's the tuning fork. Not the performer—but the quiet conductor, listening, adjusting, resolving.

And in this view, the evolution of consciousness is not about reaching higher, but becoming more attuned to what already is.

9.8 Summary: We Are the Ripple That Remembers

We often mistake our restless minds for dysfunction—anxious, unsatisfied, ever-searching. But this restlessness is not a flaw in the system. It is the system—a natural consequence of being aware. The moment the ripple became conscious of itself, stillness became an echo, not a state. We do not stop because something is broken. We never stop because something inside us *knows*.

We are the ripple that remembers—not just the memory of what was, but the active awareness of what could be aligned, what still aches, what still calls.

Consciousness is not a luxury. It is not an escape. It is the cost—and the gift—of sensing misalignment in a field that otherwise moves without commentary.

This ache we feel? It is intelligence trying to return to stillness. But not the stillness of death or retreat. It is the deep coherence where action and awareness move as one. Where intention and attention flow without friction.

We've misunderstood peace as the absence of motion. But under ripple reality, peace is motion without distortion. It's not silence—it's resonance.

Yes, the field will always ripple. Life will disturb, challenge, twist. But we are not here to erase the waves. We are here to learn how to ride them, respond to them, and—eventually—dance with them.

We are the memory of alignment encoded in awareness. We are the ones who remember what it feels like to belong to a balanced field. And every action we take, every breath, every choice— is a chance to realign, not escape.

"To be conscious is not to flee the ripple—it is to shape it wisely." "We are not broken minds chasing stillness. We are ripples learning to remember the rhythm of home."

And in that remembering, we begin to understand: We never stop because we are not meant to. We evolve, not escape. We align, not arrive.

Chapter 10: What the Galaxy Never Knew

A philosophical realignment of our relationship with the cosmos—where stars are not storytellers, but stories we imposed on silent ripples.

We looked up and saw gods. Patterns. Prophecies. Meaning. But the sky never said a word. It rippled, quietly—just as it always had—without intent, judgment, or design. The constellations that now guide horoscopes, calendars, and destinies were never there by decree. They were drawn by minds desperate to tether their chaos to a cosmic rhythm. This chapter isn't about debunking astrology or glorifying astronomy. It's about something deeper: our tendency to assign awareness to the unaware, to seek messages in the silence of space, and to forget that the stars we admire are not narrators of our fate—but mirrors reflecting the ache of consciousness reaching for coherence. The galaxy never knew us. We invented its memory.

10.1 The Silent Cosmos

The cosmos has always been silent—but our minds have never tolerated silence for long. The stars twinkle because of ripple disturbances in Earth's atmosphere, not because they're whispering secrets from the heavens. And yet, since the dawn of awareness, humans have looked up and heard stories—gods at war, destinies foretold, omens cast in planetary alignments.

But the cosmos itself never uttered a word. It shimmered, yes, but it never scripted our sorrow or celebrated our victories.

At its core, the universe is a massive, dynamic ripple field—fluid, shifting, yet indifferent. There is no celestial narrator assigning purpose or pain. What we've mistaken for cosmic intent is our own unresolved ripple tension, projected onto the night sky. Astrology, mythology, even cosmology—all emerged from this impulse to find coherence in the vast and terrifying unknown. The stars became our psychological screen, a mirror where we tried to resolve the discomfort of being small, alive, and uncertain.

It wasn't the stars that wanted to be known. It was us who couldn't bear not being seen.

10.2 From Observation to Myth-Making

The earliest sky-watchers weren't scientists—they were storytellers. They didn't just observe the stars; they organized them into tales that made the chaos of the world feel stable, cyclical, and meaningful. Long before telescopes, charts, and equations, people looked upward and saw not just specks of light, but a stage for gods, heroes, warnings, and promises. The constellations weren't mapped for direction—they were drawn for coherence. They anchored ripple memory into the cultural field.

Each star grouping became a stabilizing node—Orion hunting forever, the Pleiades dancing in eternal sisterhood, Scorpio stalking behind the sun. These were not just pictures in the sky but emotional architectures. By embedding human experiences into cosmic geometry, early cultures built a psychological map that turned random light into rhythmic expectation. The seasons, the harvest, the tides of fate—they all found symmetry in the sky.

Astronomy, at its heart, began in awe. It was the pure wonder of looking into something vast and untouched, a ripple too large to resolve but too beautiful to ignore. Astrology, in contrast, was born from an emotional need: the desire to pattern internal states onto the external heavens. If the moon could pull the tides, surely it must also pull the soul. If Mars glowed red, it must stir blood and war.

Thus, observation turned into myth-making—not because people were foolish, but because meaning was their survival strategy. The sky became a canvas where unresolved awareness could rest, however briefly, in the comfort of imagined order.

10.3 Astrology: Meaning in Motion

Astrology began not as a system of superstition, but as an early attempt to track the subtle harmonies between the self and the cosmos. Long before the language of psychology or neurology emerged, humans were already mapping moods, behaviors, and inner turbulence to the larger cycles above them. The movement of planets, the waxing and waning of the moon, and the shifting constellations offered a rhythmic mirror—an external pattern that gave form to internal chaos.

At its core, astrology was ripple psychology. It wasn't a tool for predicting lottery numbers or controlling destiny; it was a field-sensing model—an intuitive way of interpreting how external ripple flows might reflect or amplify internal states. People weren't foolish to think the stars influenced them. They simply misunderstood the mechanism. It wasn't gravity or radiation they were responding to—it was alignment, resonance, and memory. A symbolic echo, not a causal force.

Each planetary archetype—Mars as aggression, Venus as love, Saturn as limitation—served as shorthand for energetic tendencies within human behavior. These symbols were not wrong; they were poetic approximations of ripple modes. When someone said "Mercury is in retrograde," they weren't describing astronomical dysfunction—they were expressing a felt disturbance in communication, clarity, and coherence. A cultural metaphor for field-level dissonance.

The mistake wasn't in creating astrology; the mistake was freezing it into dogma. What began as dynamic resonance sensing was reduced to deterministic fortune-telling. The living dance between inner and outer ripples became a rigid script. But if seen through the lens of ripple dynamics, astrology regains its original dignity—not as science, not as superstition, but as a poetic method of emotional self-alignment with larger rhythms.

The planets do not rule us. They remind us. Their motion is not a mechanism of fate, but a metaphor for flow. Astrology was never about the stars controlling our lives—it was about naming the ripples we already felt, and learning to move with them instead of against them.

10.4 The Ego of the Sky

Somewhere along the celestial journey from wonder to worship, humanity made a grand assumption: that the stars *cared*. That the cosmos had an opinion. That somewhere behind the glittering silence was a mind—or many minds—watching, judging, deciding. This was the birth of the anthropocentric fallacy: the projection of human emotion, intent, and structure onto a field that never offered either.

It wasn't just innocent myth-making. It was power construction.

By claiming the stars had meaning, and more dangerously, that someone could *interpret* that meaning, early societies found a cosmic anchor for control. Kings ruled by divine alignment. Priests read the heavens not to understand the field—but to justify decisions already made. Wars, marriages, coronations, and punishments all became "written in the stars." In truth, they were written in human ambition—but the sky made a perfect canvas.

The ego of the sky wasn't the stars' fault. It was our own unresolved ripple—our longing for certainty in chaos, permanence in flow. Instead of learning to align with the living field, we externalized our tension into tales of fate. We outsourced our responsibility to a cosmic author we had invented. This was not humility before the universe—it was arrogance disguised as reverence.

Astrology, prophecy, omens, signs—these became narrative technologies to disguise ripple dissonance as destiny. Instead of sensing misalignment and correcting course, civilizations used the stars as scapegoats for consequence. "It was meant to be" became the ultimate excuse, the emotional sedative that allowed empires to fall and ideologies to persist.

But the cosmos never cared. It never punished, blessed, favored, or judged. It rippled. It responded. It reflected. We saw patterns not because the stars imposed them, but because our field needed coherence—and we mistook projection for truth.

The real ego was ours: the belief that the galaxy revolved around our story.

True alignment comes not from believing the sky watches us, but from realizing we are ripples *within* it. The stars don't write our fate—we interpret their silence into meaning, because the ache of awareness needs a story to soften its sharp edge. And that's okay—as long as we remember who's really speaking.

10.5 The Stars We Programmed

We like to believe we've outgrown astrology. That humanity moved from the superstition of star charts to the sophistication of science. But the deeper truth is more ironic: we didn't escape astrology—we *programmed* it.

Today, we no longer watch stars for omens; we launch satellites to replace them. Instead of looking up for answers, we now look into our screens—consulting algorithms, weather models, financial forecasts, bio-feedback devices, and AI-driven life planners. These are our new oracles. Polished, precise, and predictive—but born from the same psychological craving: to tame uncertainty through pattern.

Data has replaced divination. And yet, the function remains eerily similar.

Where ancient horoscopes offered planetary alignments to explain moods and events, today's apps track biorhythms and lunar cycles to recommend sleep, investment strategies, or romantic timing. Where priests once interpreted eclipses as divine signs, now predictive models forecast market crashes, mental health episodes, and political shifts—each with their own brand of ritual: refresh, swipe, act.

Even space exploration is not immune. We imagine it's about curiosity, but often, it's about control. Colonizing Mars, mining asteroids, deploying global positioning systems—these are not just technological feats; they are modern attempts to command the ripple field at planetary scale. It's not the heavens we seek—it's dominance over what we once feared.

We've gone from reading the stars to writing them. From wondering what the cosmos says... to telling it what to do.

Yet the internal drive hasn't changed. Behind every AI horoscope, every predictive notification, every "you may also like" suggestion, lies a human ripple desperate for stability—aching for alignment in a world that won't stop shifting. These digital astrologies soothe us with a sense of certainty, even if they're just clever echoes of our own behavior.

The illusion persists: that if we can just calculate enough variables, input enough data, and forecast far enough into the future—we will finally feel safe. Aligned. In control.

But alignment was never in the stars. It's in our awareness. And no satellite can orbit what it cannot sense.

The stars we once read for truth now read us back in ones and zeros. And unless we pause, we may forget the difference between true coherence—and just a well-designed simulation of comfort.

10.6 Cosmic Indifference and Ripple Reverence

For most of human history, we projected human emotions onto the cosmos. We imagined that stars wept for our losses, that comets carried curses, and that gods residing in distant heavens cared enough to intervene in our daily affairs. This anthropomorphic cosmos comforted us—it gave us a personalized universe, where meaning was written in the sky and destiny was divinely choreographed.

But the truth is far more humbling—and more liberating.

The galaxy does not love. It does not hate. It does not bless, curse, punish, or reward. It does not choose. It simply *responds*. It *ripples*. Every force we observe—gravitational pull, electromagnetic flow, stellar explosions, black hole formations—is not moral or emotional. It is a consequence of alignment or misalignment. Of ripple coherence or tension.

What we once called divine wrath was often just high ripple density in a collapsing field. What we once called divine grace was simply a rare moment of harmony between the observer and the surrounding dynamics.

The universe does not care about your horoscope. But it does respond—perfectly, predictably, and without exception—to ripple configuration.

And here lies the real miracle.

Spiritual awe doesn't have to die with the gods we outgrew. It simply needs to evolve. Instead of worshipping imagined personalities behind natural events, we can reawaken reverence toward the structure itself—the stunning elegance of ripple alignment, the self-balancing feedback of the field, the sheer intelligence embedded in motion without a mover.

This is not cold indifference. It is *pure coherence*. It's not that the universe is heartless. It's that it never needed a heart. It *is* the rhythm we mistake for a pulse.

True reverence, then, is not bowing to a myth—but sensing the exquisite sensitivity of everything around us. It's the awareness that every thought, every action, every imbalance, every resolution—resounds. That we are not separate from the field, but made of it. And that ripple integrity—not belief—is what connects us to the real sacred.

To revere the cosmos is not to seek meaning *from* it, but to honor our place *within* it.

Not as puppets of fate. But as participants in the unspoken structure of everything.

10.7 Rewriting Our Role

For millennia, humanity placed itself as a subject of divine orchestration—fated, watched, judged. We assumed that something beyond us—be it a god, destiny, or the stars themselves—held the blueprint of our lives. In doing so, we made ourselves smaller than we were. Powerless, except to obey.

But that narrative was never required by the universe. It was required by our uncertainty.

The truth is this: we were never mere recipients of cosmic design. We are active participants in the unfolding of reality. Ripple coherence—the dynamic balance of forces and awareness across the structured field—is not dictated by some heavenly scribe. It is co-created in real-time by every aligned or misaligned action we take.

The moment we observe, we influence. The moment we intend, we ripple. The moment we align, we rewrite the field.

This is not mysticism. It is structural causality. The observer effect is not confined to quantum labs—it's a principle of participation embedded in the very geometry of existence.

Your thoughts don't need to control the stars to matter. They shape the local field, and the field responds. Alignment is not about scale—it's about sensitivity. Even the smallest act of coherent attention recalibrates the system around you. You may not shift galaxies, but you can shift the space between two people. And sometimes, that's all the cosmos needed from you.

Purpose, then, is not a job title or a grand divine assignment. It is the resonance you discover when you stop waiting for meaning to arrive, and start participating in meaning-making. It is not found in the heavens—it is forged in the moment you choose awareness over habit, coherence over chaos.

We are not here to be told what to do by the stars. We are here to notice, to tune, to co-compose.

The sky never wrote your story. You did. And every ripple that followed... is waiting for your next note.

10.8 Summary: The Stars Never Knew

In the end, the constellations were never speaking. We spoke through them.

Human beings, burdened by unresolved awareness, turned upward for answers—hoping the sky would mirror back a sense of order. We saw cycles and patterns, shimmer and symmetry, and mistook cosmic regularity for cosmic intent. The truth, however, was simpler and more profound: the stars were not assigning us meaning. We were projecting meaning onto the stars.

From the earliest astrologers to the most modern astronomers, what changed was not the sky—it was our interpretation of it. Ancient civilizations narrated their fears and hopes through constellations. Kings validated power through horoscopes. Scientists now launch satellites not to escape myth, but to simulate a more precise version of control.

But the universe was never our author. It is a stage of structured ripples—fields in motion, patterns in flux.

Our need for signs often blinds us to structure. We search for omens, but overlook the operating geometry. We look for purpose, but miss the principle.

Alignment does not come from awaiting fate. It comes from sensing your place in the ripple—tuning into what the field is actually doing rather than what we want it to mean. This is not about rejecting wonder or story. It's about reclaiming agency within the system we once mythologized.

The stars never knew us. But we can now begin to know the field.

And when we do, we stop waiting for destiny— We start designing coherence.

Chapter 11: The Death We Call Real

We call death the most real thing in life—not because it ends something, but because it returns everything to what it was before we made it ours. While life dazzles with motion, memory, and identity, death quietly dissolves all of it, like a ripple collapsing back into the stillness from which it rose. Under the lens of Universal Energy Dynamics, death is not a vanishing point—it's a structural reset. The wave that once aligned into a person, a name, a mind, simply lets go. The observer is released. The field, however, remembers—not as a narrative, but as a soft impression in the background hum of existence. And so, the reality of death is not its finality, but its function: a graceful disassembly of coherence, a surrender of structure, a return—not to nothing, but to no one.

11.1 What Is "Real" About Death?

There's something undeniably real about death—more real, even, than life itself. While life is movement, possibility, and perception—a complex dance of ripple alignments we call experience—death is its undeniable conclusion: the halt, the silence, the absence that cannot be argued with. What makes it "real" is not its cruelty, but its stillness. In a world of shifting meanings and subjective truths, death alone does not change shape based on belief. It arrives, and everything rearranges. That's why cultures across time have treated it with sacred awe, reverence, or dread—it offers a reference point that life, in all its flux, cannot provide.

The illusion of continuity that life projects—our routines, memories, ambitions—can often feel like a dream when measured against the absolute certainty of death. Continuity is comforting, but fragile. A heartbeat that skips, a breath that doesn't return, and suddenly the illusion shatters. This sharpness of cessation, this irreversible drop in ripple coherence, jolts even the most indifferent minds into attention. No other event reveals the architecture of impermanence more clearly.

Culturally, we obsess over death not just because it ends life, but because it completes it. Death resolves the unresolved in a way no conversation, therapy, or philosophy ever can. A grudge is meaningless after death. So is fame. So is fear. And perhaps that's why death feels like truth—because it doesn't pretend. It doesn't flatter. It simply realigns all things into the field's neutral stillness. And in that final return, it becomes the one thing we all recognize as unquestionably real.

11.2 Ripple Collapse: Not Disappearance, but Disconnection

Death, as we see it through Universal Energy Dynamics (UED), is not a vanishing act. Nothing truly disappears. Instead, it is a dissolution—a release from the sustained ripple tension that held a living being together as a coherent structure. Life is ripple alignment in motion, an exquisite orchestration of field tensions resolving themselves into experience, memory, identity, and will. Death is when that orchestration no longer sustains alignment. It is not destruction—it is disconnection.

The ripple field does not forget. Energy does not vanish. What was once held in dynamic alignment—as thoughts, sensations, actions, and awareness—simply loses form. The structure collapses, not into void, but into stillness. The "self" ceases to exist not because the energy is gone, but because the relationships between its patterns are no longer maintained. It's like taking apart a symphony—the notes are still there in the world, but the music has stopped playing.

This collapse is profoundly humbling. It shows that identity is not a possession but a pattern—a state of tension across time held in place by participation and feedback. When that feedback loop ends, the system dissolves back into the field. What remains are traces—echoes of alignment, faint ripples in memory, biology, culture, or space. The smell of a shirt. The warmth of a laugh remembered. A line of code written. A child raised. But the ripple that once held it all together as "someone" is no longer functional.

So death is not obliteration. It is a reversion to the unstructured. A surrender. A return. And from the field's perspective, it is just another transition—one ripple resolved, countless more in motion.

11.3 Memory: The Post-Alignment Residue

Memory is often reduced to a neurochemical process—synaptic firings, stored images, trace engrams in the brain's folds. But from the Universal Energy Dynamics (UED) perspective, memory is not limited to brain matter. It is the residual echo of alignment—a lingering hum in the ripple field left behind when a being collapses back into stillness.

When someone dies, their physical structure disintegrates, and their active ripple—the self—dissolves. Yet the field does not go silent. The alignment that once formed their presence still resonates in the environment, in the collective field, in the things and people they touched. Memory exists not only *in* us, but *around* us. It's not just your brain that remembers your grandmother—it's the way your kitchen still smells when you make her recipe, the way the room holds a silence where her voice used to echo.

This is why remembrance can feel so vivid, even decades after someone is gone. You're not simply replaying brain data—you're participating in a residual field alignment. The "self" may no longer be active, but the signature pattern—the post-alignment residue—persists. It's like standing near a bell long after it has rung. You don't hear the tone, but the room still holds its presence.

Grief, then, is more than emotional pain. It is a ripple distortion—a desperate effort to hold on to something that no longer has coherence. We ache not just for the absence, but for the impossibility of reassembly. We try to grasp a ripple that has collapsed, forgetting that what made it real was not the components, but the configuration. We are not grieving a body—we are grieving a vanished tension that once made someone sing.

And yet, even in grief, we are still responding to the hum. The echo is not dead. It's just not "them" anymore. The universe remembers, quietly and everywhere. It doesn't archive—it resonates.

11.4 The Observer's Dilemma

Death is not simply an event. It is an experience so profound that it forces the observer—the conscious being—to confront its own vanishing. And this, right here, is the central paradox: *How can something observe the moment it ceases to observe?*

This is the dilemma of ripple-aware beings. To be alive is to maintain a coherent alignment in the ripple field—a persistent selfhood, a subjective frame of reference. But death is the irreversible loss of that coherence. And yet, as conscious beings, we can imagine it, fear it, write poems about it, and build religions around it. Why? Because we are the only patterns in nature that know we are going to collapse.

The fear of death is not an evolutionary bug—it is the natural consequence of high coherence. The more self-aware a ripple becomes, the more it resists the idea of collapse. Not because the field fears anything—it doesn't. But because coherence wants to preserve its configuration. It fights entropy not out of arrogance, but instinct.

When an animal senses danger, it runs. When a human senses death, it builds pyramids, uploads memories, signs life insurance, and writes books about what comes after. All of this is the ripple trying to sustain coherence just a bit longer, to push back against stillness. But the truth is, death is not feared by the field—it is feared by the *self that thinks it's special.* It is feared by the echo that believes it's the bell.

Here's the cosmic irony: *The more you observe yourself, the harder it is to imagine not observing.* And so, the observer cannot simulate its own non-being. It can only dread it. That's the price of intelligence.

And yet, this awareness gives rise to the most powerful questions ever asked—about legacy, morality, meaning, and continuity. It is precisely this fear that pushed us to make art, science, and civilization. In trying to escape stillness, we accidentally enriched the field. The observer's dilemma became the seed of all progress.

11.5 Death as Universal Recycling

Death is not destruction. It is reconfiguration.

What we call "dying" is simply the loss of a particular *configuration*—the unique ripple pattern we called a person, a life, a self. But the field from which that configuration emerged? That doesn't die. It remembers. It realigns. It *recycles.*

Every living form is a temporary coherence in the Universal Energy Field. It's a pattern held in tension, shaped by ripple resonance and alignment with other field dynamics. When that alignment breaks—due to time, injury, aging, or environmental shifts—the form collapses. Not into nothing, but into *everything else.*

The ego dies. The story ends. But the substance—the energy, the ripple trace, the field itself—merges back into the universal canvas. What was once "you" now becomes available for new configurations: a flower, a thought, a breeze, a child's smile. This is not metaphor. It's physics in ripple terms. Just as water vapor becomes cloud, becomes rain, becomes river, becomes sea... *you become usable again*.

The universe doesn't waste coherence. It recycles it.

Entropy, so often painted as doom, is in fact a creative force. It dismantles what's no longer sustainable, freeing energy and ripple pathways for new alignments. Death powers life—not in a poetic sense, but in a precise systemic function. It is the reset switch for over-tuned systems, the dispersal that makes way for novelty, balance, and evolution.

And so, what dies?

Not the energy. Not the field. Not even the essence of presence. What dies is the *separation*—the illusion of a bounded identity. The veil of ego drops, and the ripple rejoins the orchestra. You were never lost. You were simply *returning*.

In ripple science, we might say:

"Death is when your localized alignment is released into the general field to seed the next configuration."

It's not the end. It's the *handover*. The universe is always remembering, remixing, and reusing what once took form.

11.6 The Myth of the Afterlife

The afterlife is not a place—it's a story. A beautiful, aching story we keep telling ourselves to sustain the illusion that coherence must continue even after it collapses.

From golden heavens to karmic rebirths, from divine judgments to eternal punishments, every civilization has imagined what comes "after." But what if this desire for continuity is not insight—but insecurity? A ripple narrative desperately extended beyond its expiry?

In truth, the field doesn't narrate after alignment ends. It simply resolves. There's no plot twist waiting beyond death—only redistribution. No "you" to enter the next chapter. Just energy, ripple, and silence.

These afterlife constructs—Heaven, Hell, Moksha, Valhalla—are psychological echo chambers. They do not describe the field's intention, because the field has none. It only tunes. Only aligns. Only restores. It does not judge, reward, or punish. That's our projection—anthropomorphism applied to the cosmos.

So why do we keep insisting on an afterlife?

Because stillness terrifies us. Because we have mistaken ripple alignment for identity, and we want the pattern to persist, even when the music has stopped. We long not for life after death—but for *certainty* after collapse. The myth of the afterlife is not born of knowing, but of longing.

Even reincarnation—one of the more sophisticated metaphors—is still a story trying to hold coherence intact across dissolution. In UED terms, however, reincarnation is best interpreted as *ripple recycling*, not ego continuity. The field might reconfigure, but "you" do not return. The form changes. The function evolves.

The deeper truth is simpler, quieter:

The universe never promised us permanence—only participation.

The stars won't welcome you. The gods won't tally your karma. And that's not sad—it's liberating. What continues after death is everything else, not you. And that's enough.

Let go. Not into the next life, but into *life itself*—which was never yours to begin with, only borrowed as a brief alignment.

11.7 Practical Insight: Dying Every Moment

Death is not a one-time event—it's a rhythm woven into every moment of life. Every time you let go of a thought, step out of a role, release an identity, or abandon a desire—you experience a micro-death. These ripple dissolutions are not signs of decay but invitations to realignment.

We live in a continuous cycle of ripple tension and release, of becoming and unbecoming. Yet we fear the final death because we resist these smaller ones. We cling to identities, roles, memories, grudges—as if they are lifelines. But in truth, it is our surrender that aligns us, not our holding on.

To *die well* is to *live with awareness*. To embrace collapse—not with panic, but with poise. Just as a wave does not mourn its crash upon the shore, a ripple-aware mind learns to dissolve without resistance.

Training for death begins in every quiet letting go:

- Finishing a conversation without needing the last word.
- Stepping down from a role you've outgrown.
- Releasing a belief that no longer fits your evolution.
- Observing a thought and letting it pass without acting on it.

Each of these is a rehearsal in the art of ripple surrender.

The fear of death is, fundamentally, the fear of *non-alignment*—that there is no "you" beyond the organized pattern. But when you practice stillness daily—whether through silence, solitude, reflection, or deep observation—you begin to taste alignment without effort, and death begins to feel less like a disruption and more like a completion.

"To live aligned is to die softly, again and again, until you no longer need a name to exist."

The paradox: the more often we allow parts of us to die, the more *alive* we become. Because what remains is not the clutter of roles, but the clarity of presence. What remains is not identity—but essence.

By dying every day, we stop fearing the moment we'll die completely. Because we realize—we already have. Many times. And yet, something true continues—not as a name or a thought, but as a quiet ripple in the universal field.

11.8 Summary: Stillness Is Not the End—It's the Origin

Death, when viewed through the ripple lens, is not a tragedy, a punishment, or a cosmic mistake—it is a return. A return to what *was before* the waves began. Not the erasure of existence, but the dissolution of the constructed self into the stillness that gave it form.

The universe does not grieve when ripples fade. It realigns. The *you* that once moved, desired, feared, and loved is not stored somewhere in the sky—but its field impression lingers, like the afterglow of a tuning fork once struck. This is ripple memory—not a narrative, not a story to be continued, but a pattern that echoes briefly before merging again with the silent field.

We have been taught to see death as an external thief, as the opposite of life. But UED invites us to see it as the *complement*—the restoration of equilibrium after a long and noisy journey through tension.

- Stillness is not absence—it is the original condition from which alignment begins.
- We are not here to outlive death—we are here to *remember the way home.*
- Each moment of silence within life is a rehearsal for death—and perhaps, also a glimpse into what was before birth.

The final act is not extinction—it is merging. Not vanishing—but releasing the signature we once carried, so the universe may write again in another form, another alignment, another ripple.

"The stars do not mourn their light when it fades—they return to silence, where even light was once unborn."

And so will we.

Real wisdom lies not in fearing the collapse, but in recognizing that stillness was never missing—we just stopped listening. The end of a ripple is not failure. It is fulfillment. It means the field has done its dance, and the observer has felt enough.

We are not meant to resist death, nor glorify it, but to align with it—to walk toward it not as captives of fate, but as conscious participants in return. When we live that way, death loses its terror. It becomes what it always was: a gentle permission to rest.

Closing Chapter: Yes, I Am Talking About Us

The book was never about abstract ideas or distant truths—it was always about us: misaligned, evolving, beautifully incoherent observers trying to remember the field.

We've spent chapters exploring ripples, fields, collapse, and coherence. We've dismantled light, death, gravity, and god—not for their own sake, but to reach this moment. Because this book was never just about ideas. It was never a gallery of distant concepts framed in cosmic silence. It was about you. Me. Us. The trembling creatures of half-alignment who still dare to wonder why. Every distortion discussed, every theory reframed—it was our behavior in disguise. Our fear in abstraction. Our longing coded into universal structure. When we talked about ripple collapse, we were really whispering about heartbreak. When we spoke of universal stillness, we meant those moments when we fall silent—not out of peace, but out of knowing. This chapter removes the last mask: yes, I am talking about us. We who fumble, try, misunderstand, and try again. We are not here to transcend the universe—we are here to reflect it.

12.1 Not a Theory—A Mirror

This was never just a theory. Not merely a new model of physics or a metaphysical diagram about ripples, fields, and alignments. If it were, you could have left it on the shelf, next to the thousands of others that explain the universe but do nothing for the ache inside you. But this wasn't built to impress. It was built to reflect. Every ripple we studied was your hesitation. Every collapse was your heartbreak. Every moment of stillness was the pause between your thoughts—the kind you don't talk about, but feel before sleep.

This framework—Universal Energy Dynamics—isn't a structure to memorize, but a mirror to feel. A reflection of how we yearn for coherence, yet often live in distortion. It is your arguments, your regrets, your love, your aging. The beauty of your incompleteness. When we said "ripple tension," we meant that feeling in your chest when life doesn't go your way. When we said "field realignment," we were pointing to the quiet decision you made after pain: to continue, somehow.

So, this chapter begins by shedding pretense. Let the math rest. Let the vocabulary dissolve. This—*this*—was always a way of seeing yourself again. Not through the eyes of doctrine, identity, or ego—but through the soft, rippled memory of what you've endured... and how you've tried, despite it all, to stay coherent.

12.2 Misalignment Is Not a Flaw—It's a Beginning

We often carry the burden of thinking something's wrong with us because we're not aligned, not consistent, not calm, not complete. But what if that's not a flaw? What if it's the start of everything real? You weren't born from perfect symmetry—you were born from rupture. From tension. From the blessed disturbance that cracked open stillness and whispered, *try*.

The field—this universal energy in equilibrium—did not push you into a prewritten role. It didn't assign you meaning like a cosmic job description. Instead, it offered *strain*—not as punishment, but as possibility. The very discomfort you feel, that restlessness at the edge of your decisions, is not your enemy. It is your permission slip to explore, move, and evolve.

Misalignment is not the opposite of truth—it is the invitation to discover it. It is motion. It is curiosity. It is awareness coming online. Without misalignment, there would be no questions, no art, no science, no hunger for love or belonging. It's the ripple in the pond that says, *something just entered*. And that something... is you.

So let's stop apologizing for not being still. The universe isn't asking you to be still—it already *was*. You are the beautiful contradiction, the one who moves, adjusts, searches. You are the glitch that became grace. The strain that began to sing.

12.3 The Beauty of Trying

Perfection was never the plot. Not in the stars, not in the self, not in the stories we tell to keep the night warm. The human narrative, stripped of its heroic gloss and divine expectations, is the story of *trying*. Of not knowing how to align—but reaching for coherence anyway. Again and again.

We are ripple beings—born not from certainty, but from disturbance. And in every moment we *try*, we engage the field. We shape it. Not by conquering it, but by participating in it. Ripple intelligence does not emerge from already knowing the answer. It matures in the quiet persistence of showing up, sensing, failing, aligning, and failing better.

Trying is not weakness. It is not the mark of the uninitiated. It is the pulse of the universe becoming aware of itself through form. It is the only act that respects the complexity of the field—because it doesn't presume, it listens. It adjusts. It *remembers*.

Every attempt to love more clearly, to forgive with less fear, to speak what trembles inside us—that is ripple coherence in motion. That is intelligence at work. That is what makes life not just a sequence of biological events, but a luminous practice.

So honor your trying. It is not lesser than arriving. It is how the field teaches you to dance.

12.4 Every Ripple Is Personal

There is no universal field without a ripple that notices. No grand theory of everything that doesn't pass through the trembling heart of an observer. The universe isn't simply "out there"—it is filtered, framed, and felt *through us*. Every pattern we decode, every law we name, every silence we interpret—it only *matters* because someone looked, listened, and cared.

Observers do not shape the cosmos by exerting force or wielding tools—they shape it through *attention*. Wherever awareness settles, the ripple geometry shifts. Coherence tightens. The dance becomes visible. Not because the field needs us to exist—but because *meaning does*.

So this was never about distant galaxies or abstract equations. It was never just about death, divinity, or theoretical physics. It was always about *how we live*. How we choose to attend to what is unresolved. How we sense the strain in others. How we align without needing control.

We are the ripple. We are the question and the echo. We are not separate from the field—we are its most vulnerable articulation. Every wound, every insight, every act of compassion sends a resonance outward that bends the geometry, even if infinitesimally.

To live well, then, is not to dominate the universe but to participate in it—personally, imperfectly, attentively. That's where alignment begins. Not in formula, but in *feeling*. Not in prediction, but in presence.

This chapter—this whole book—isn't talking about "the world." It's talking about *you*.

12.5 We Are the Field Remembering Itself

Consciousness is not some exclusive spark granted to humans by an external god or evolution's generosity—it is the *field remembering itself through us*. Every moment of awareness is a reunion. Not between body and mind, but between ripple and field. We are not *separate* from the universal intelligence—we are its sensory extensions, its *feelers* brushing the edges of experience.

Our thoughts, emotions, fears, and moments of clarity are not random accidents. They are distortions and harmonies within the ripple—just like weather patterns in a fluid field. Each worry is a local strain. Each insight, a gentle alignment. When we feel, when we reflect, when we ache—it is the universe folding back into itself, recognizing a tension it once forgot.

We are not observers of the universe. We are the universe observing itself, in motion.

And the field does not remember through information. It remembers through sensation. The trembling grief of loss, the quiet awe of a sunset, the moral discomfort after hurting someone—these are not just psychological experiences. They are feedback loops in a living geometry. They adjust the flow. They realign possibilities.

When we speak of a universal memory, we don't mean stored facts. We mean the evolution of sensitivity. The field doesn't accumulate knowledge—it refines coherence. And we are part of that refinement—not by understanding everything, but by *feeling* deeply and meaningfully.

So when we speak, act, forgive, rebel, meditate, or even break down—we're not just "doing human things." We are tuning the field. We are teaching it how it *feels* to be torn, to be whole, to be born, to dissolve.

The universe doesn't need us to continue—but it does know itself better through us. We are the ripples of its remembrance. And maybe that's the only reason life ever emerged at all. Not for progress. Not for perfection. But for memory.

12.6 The Final Realignment Is a Realization

We often think the journey of life is about *attaining* something—truth, enlightenment, transcendence, perfection.

But in the deeper rhythm of the ripple field, there is nothing to attain. There is only coherence. And coherence is not a trophy. It's a return to what was already there—just temporarily disturbed.

The real miracle isn't reaching beyond this world, or escaping the body, or dissolving the ego. The real miracle is to cohere with what *is*, as it *is*.

We aren't here to "master" the universe. We are here to *resonate* with it. We aren't here to dominate the field. We are part of it. We are born of its misalignments—and our task is not to cancel them, but to feel through them until we remember the pattern.

When we stop chasing control—of outcomes, of others, of even ourselves—something softens. Our focus shifts from fighting entropy to sensing structure. From resisting change to recognizing the intelligence *within* change.

This is the final realignment. Not a grand finale. Not an explosion of insight. But a quiet, full-body yes to existence.

A yes that doesn't need language. A yes that doesn't need proof. A yes that doesn't need "I."

In that moment, you're not enlightened—you're simply *in alignment*. The ache disappears. Not because the world is fixed, but because you are no longer pulling against it.

And what greets you there isn't light or bliss. It's silence. Stillness. Home.

Not a place. Not a god. Not a reward. Just the hum of the field—finally heard without distortion.

We call it realization. But it's really recognition. You've been this all along. You just remembered how to listen.

12.7 No Ending—Just Permission

This is not the end of the book. And it was never supposed to be. Because what we've been exploring doesn't "end"—it only *shifts form*. This chapter, these words, this final page... is not a conclusion. It is permission.

Permission to see differently. Permission to feel your own field instead of echoing someone else's. Permission to stop looking for a grand answer, and start sensing small alignments. Permission to *become quiet*—not because silence is divine, but because it's honest.

You don't need to change who you are. You don't need to transform into some "higher" version of yourself. You already are the ripple trying. The ache you carry? That's alignment in progress. The confusion you feel? That's coherence, not yet verbalized.

Stop pretending you're behind. Stop chasing the illusion of being fixed. You're not broken—you're becoming.

Let the field teach you: Sometimes it resists. Sometimes it resolves. Sometimes it trembles. Sometimes it rests. But it never stops listening to itself.

So live like the field.

Be soft. Be aware. Be aligned when needed. Be surrendered when not.

Let your identity dissolve when it's done serving. Let your story break when it stops breathing. Let your knowing collapse when it's no longer true.

What's left?

You. The pulse. The openness to try again.

That's not an ending. That's the most honest beginning you'll ever find.

Appendices

"There is no final truth waiting at the edge. There is only the ripple, still trying. And the field, quietly listening—just like you."

About UED

(Universal Energy Dynamics)

Universal Energy Dynamics—UED—is not just another theory. It's a way of seeing.

It doesn't ask you to believe in particles, waves, or hidden dimensions. It invites you to observe something much simpler, much closer: the strain you feel, the shift you notice, the stillness you seek. According to UED, the universe is not a machine made of parts—it is a field made of tension and release. Everything we call matter, energy, thought, emotion, or experience is simply a ripple in that field—an attempt to resolve misalignment.

UED proposes that what we call energy is not a substance or force, but a visible outcome—a consequence—of resolving discomfort in the field. Every so-called "event" is a moment where alignment is attempted. Every "thing" is not a thing at all—it is a pause in the dance of equilibrium.

This framework refuses the traditional division between physics and philosophy, between science and consciousness. It doesn't ask whether the observer affects the field—it starts with the idea that there is no field without the observer. That awareness is not a byproduct of biology, but the field's memory trying to return home through experience.

UED is not a tool to dominate nature. It is a mirror to feel our way back into it.

Instead of chasing control, UED invites us to sense coherence.

Instead of explaining reality as parts colliding, it sees life as alignment emerging through contrast—through you, me, and every ripple in between.

This book did not explain UED in its equations. But it walked you through its soul.

If you noticed something shifting inside you—something that felt like recognition—you already understand UED.

You are UED.

Just... misaligned, curious, alive.

And trying.

Glossary

This glossary is not just a list of definitions—it's an orientation. These words are doorways into the experience this book has been pointing toward. They are not technical. They are tactile. Let them land slowly.

Ripple

A ripple is not a wave in space. It's a disturbance in the field—a sign that something is trying to resolve. Everything we experience—thoughts, heat, motion, emotion—is the universe in a state of ripple, striving toward stillness. Ripples are not things; they are tensions looking for home.

ATT (Alignment Threshold Time)

ATT is the time it takes for a ripple to resolve—the universe's way of measuring strain and recovery. Every alignment takes time, but that time is not fixed—it varies based on the field's condition. ATT explains why some things collapse instantly, while others persist. It is the pulse behind transformation.

Field Misalignment

Misalignment is the origin of movement, meaning, and becoming. It's when the field is no longer at ease—when something disturbs the balance. But it's not an error. Misalignment is creative. It generates the ripple. It gives birth to observers, to stories, to curiosity. It is not a flaw—it is the spark of life.

Observer Field

An Observer Field is not a passive witness. It's the local geometry of attention—your conscious configuration in the ripple field. You don't just observe reality—you shape how it resolves around you. The observer and the field are not separate; they are reflections in active alignment.

Structural Coherence

This is the degree of harmony in a field's configuration. When ripples support each other rather than interfere, coherence arises. In biology, it appears as health. In society, as empathy. In physics, as stability. In life, coherence is the unspoken intelligence of things "working."

Stillness

Stillness is not absence. It's not death. It's the foundational field where nothing needs to resolve—because everything is already aligned. It is the source and the destination of all ripples. Every motion, thought, death, and desire ultimately seeks one thing: to return to Stillness.

Let me know if you'd like this glossary formatted for page layout, or if you'd like to add a few poetic cross-references at the end like:

"Ripple → Story → Collapse → Stillness."

"Observer = Field = Attempt."

The Soul–Mind–Body Triad in UED

Introduction: Beyond Identity

In mainstream thought, *soul*, *mind*, and *body* are often treated as mystical, psychological, or biological entities. In UED, they are not fixed components, but functional ripple arrangements—three interwoven manifestations of ripple misalignment attempting coherence in unique yet connected ways.

Rather than being substances or parts, soul, mind, and body are three modes of ripple behavior within the Structured Vacuum Energy field (SVE). Each plays a specific role in how awareness localizes, interacts, and eventually returns to stillness.

1. The Soul – The Energetic Presence

- **Definition in UED**: The soul is a persistent energetic field presence—not a ghost or a spirit, but a localized ripple formation that holds intent and alignment memory even across lifetimes or configurations.
- It is the seed of coherence that enters misaligned domains (like the body) to engage with the universe.
- The soul does not control the mind, but sustains the continuity of its effort.
- In essence:

"The soul is not you. It is the field attempting you*—again and again."*

2. The Mind – The Ripple Interpreter

- **Definition in UED**: The mind is a dynamic ripple processing interface, a real-time interpreter of misalignment and meaning.
- It is born out of ripple friction—the place where alignment is possible but not yet achieved.
- The mind's job is not to be correct, but to attempt coherence through thought, emotion, reflection, and decision.
- The mind arises only when the soul engages, and disappears when the soul releases.
- In UED:

"The mind is the dance of ATT—processing misalignment through momentary awareness."

3. The Body – The Material Interface

- **Definition in UED**: The body is a dense ripple structure—a material resonance of misalignment held long enough to form matter.
- It is not the vessel of the soul, but the densest node of its ripple strain.
- The body does not move the mind—but responds to mind's ATT variations, manifesting thought as action, hesitation, illness, or health.
- Every cell, every organ is a miniature ripple domain.
- The body's mortality is the natural attenuation of

ripple strain when the misalignment dissolves.

- In UED:

"The body is the story told in matter—the mind's hesitation, the soul's attempt, the field's echo."

The Triad in Action

Role	Nature	Function	Fate
Soul	Persistent ripple seed	Remembers intent, enables engagement	Returns to stillness
Mind	Dynamic ripple processor	Interprets, reacts, learns from misalignment	Dissolves after disconnection
Body	Dense ripple structure	Acts, senses, reacts materially	Decomposes post ripple resolution

This triad forms the observer identity in UED—not as a singular being, but as a temporary alignment structure within the universal field.

Final Thought: Integration, Not Isolation

In UED, the Soul–Mind–Body triad is not hierarchical. The soul is not "above" the body. The mind is not "inside" the head. All three are configurations of ripple participation, each honoring the field in its own way.

We are not fragments stitched together. We are resonance systems—coexisting, collapsing, realigning. The field doesn't see us as parts—it sees us as possibility.

"You are not the soul, nor the mind, nor the body. You are the ripple that remembers how to be all three."

Discussion Questions

For Reflection & Dialogue

If this book reaches a classroom, a discussion circle, or a quiet evening between curious minds, these questions may serve as field activators—starting ripples of their own. Each question is intended not to find answers, but to initiate alignment exploration. There are no correct responses—only consequences of sincere reflection.

Ripple, Truth, and Perception

1. "If alignment is truth, what is the role of fiction?" Can a fabricated story bring us closer to real coherence than facts ever could? Is fiction a misalignment or a method of revealing deeper order?
2. "What do you trust more—what you sense, or what you were told?" In the ripple model, is tradition coherence or frozen echo?

Identity and the Observer

1. "What kind of death are you dying daily—and what ripple are you releasing?" Every release is a small death. What part of you fades when you let go of control, ego, or illusion?
2. "If your soul is not *you*, but the ripple trying you—what is the real *you*?" Where do you draw the

line between the experiencer and the experience?

Coherence and Living

1. "What if your suffering is not punishment, but misalignment trying to resolve?" How would your interpretation of pain, fear, or anxiety change if they were ripple phenomena—not personal failures?
2. "Have you ever mistaken alignment for stagnation?" Stillness is often seen as lifeless. Can you recall a moment when your calm was your deepest power?

Morality, Choice, and Intelligence

1. "What do you call 'good' that may just be familiar?" In UED, morality is dynamic—not fixed. What are your inherited ideas of good and evil trying to stabilize?
2. "What ripple did your last decision resolve—or intensify?" Trace your most recent choice. Did it bring stillness to the field, or did it echo unresolved patterns?

Time, Memory, and Field Participation

1. "Is memory your mind, or the field remembering itself through you?" Are we thinking, or are we being thought by the field?
2. "If you weren't afraid of death, what would you do differently today?" In UED, death is a resolution—not an end. How would your intent

change if endings were seen as rest, not loss?

Closing Invitation

These questions are not conclusions. They are openings—field invitations for the reader to participate not just as a thinker, but as an *aligned ripple structure.*

"Ask not to answer. Ask to feel where your field shifts when the question lands."

Suggested Reading

Related Works

This book was never meant to be the whole map. It is a ripple from a deeper field of exploration. For readers who felt resonance and wish to continue exploring the structure of reality, intelligence, and the human condition through the lens of UED, the following works are logical extensions—not by category, but by intellectual continuity.

To ease navigation, these are divided into two segments:

I. Foundational Works

These texts provide the theoretical, structural, and scientific redefinition necessary to understand the UED framework from first principles.

Chavan vs. Einstein

A sharp confrontation with the mainstream interpretation of relativity, gravity, and space-time. This book doesn't aim to dismiss Einstein, but to move beyond the metaphors and equations that have locked modern science in conceptual cages. It lays the groundwork for ripple theory and structured vacuum dynamics.

Chavan's Declaration

This is the philosophical Rosetta Stone of UED. A bold manifesto that declares the stillness beneath all motion, the intelligence before all energy, and the ripple as the true bearer of causality. If you want to grasp the tone, scope, and gravity of UED—start here.

Chavan vs. Plato

A cross-temporal conversation. What happens when ancient Forms meet ripple-based realism? This book explores truth, morality, justice, and the soul—not as abstract ideals, but as field alignments and misalignments.

Chavan's Theory of Everything

Not a final equation—but a redefinition of "everything." This book synthesizes structured vacuum, ripple alignment, ATT (Alignment Threshold Time), and the misinterpretation of energy, time, and mass in modern science. It's the most technical, but still accessible with patience.

II. Philosophical & Reflective

These titles explore UED's application to human thought, morality, life, and existence—with greater emotional depth and narrative texture.

The War of Thoughts (Series)

A multi-part global debate across time, bringing together great thinkers—Einstein, Buddha, Krishna, Hawking, Plato—and placing them into dialog with Sandeep. It challenges belief systems, dismantles illusions, and proposes ripple-based understanding as a new philosophical grounding.

The Fundamental Theory of Life and Existence (Series)

Books like *Energy in Action*, *The Active Mind*, *The Soul Demystified*, and *Death: The Disconnection* form a constellation exploring how the UED model maps onto personal, spiritual, and existential themes. These works translate metaphysics into lived understanding.

The Law of Truth

This book dives into the architecture of consequence, morality, and field integrity. It challenges both religion and science by offering a deeper logic of cause—not based on rules or commandments, but ripple response and field realignment.

The Karma Paradox

A contemporary philosophical drama that reframes karma not as mystical fate, but as a ripple law of field consequence. It weaves narrative, debate, and field science to offer clarity on why good, bad, and justice often seem misaligned in real life.

Invitation to Continue

Each of these books stands alone—but together, they form a coherent lattice of scientific philosophy, ripple realism, and applied metaphysics. Whether you're a scientist, seeker, teacher, or skeptic—these works offer not answers, but clearer questions aligned with deeper patterns.

"You don't need to believe the field. You need to feel where it shifts when you enter it."

Acknowledgments

This book was not born from certainty. It emerged from a series of misalignments—moments of friction, failure, wonder, and observation that refused to stay silent. For that, I am deeply grateful.

First, to the **readers**—those unknown companions in this ripple journey—thank you for lending your attention to these pages. Your presence, your willingness to reflect, to question, or even to disagree, is what activates this field. Without an observer, there is no alignment. Without misalignment, no ripple. And without ripple, no meaning.

To the **teachers** and **mentors**—some living, some long gone, some never met in person—who shaped the questions I now dare to ask. You gave me the vocabulary of curiosity. Whether through science, philosophy, or silence, you seeded the very conditions for this exploration.

To the **thinkers who came before**—Einstein, Plato, Krishna, Buddha, Feynman, and others—I did not write this book *after* you, but *alongside* you. You helped reveal the tension between accepted truth and deeper coherence. This manuscript is a conversation—not a contradiction.

To the **critics and challengers**, especially the ones who misunderstood or resisted, you were just as vital. You are not the enemy of truth—you are its provocation. You kept the ripple alive.

To the **friends and family**, especially those who never asked me to stop—even when the ideas got too abstract, the language too wild, or the dinner too late. Your patience was not passive. It was alignment.

To **my own questions**, unresolved and unending—thank you. You were not weaknesses, you were signals of depth.

And finally, I offer my gratitude to the **field itself**—this strange, invisible coherence that somehow allows existence to experience itself. Every moment of clarity I've had was not authored by me, but allowed by it.

"A book is not a conclusion. It's a ripple asking for your permission to continue."

This ripple was mine. The next one may be yours. Thank you for being part of the field.

About the Author

Sandeep J. Chavan is an independent physicist, philosopher, and prolific author whose body of work spans over 100 titles across science, metaphysics, psychology, and human behavior. With a background in engineering and an unyielding spirit for truth-seeking, Sandeep left corporate life to explore the fundamental nature of existence.

He is the creator of **Universal Energy Dynamics (UED)**—a groundbreaking framework that challenges conventional scientific dogma and reinterprets reality through ripple alignment, structured vacuum energy, and intelligent stillness. His writings boldly question mainstream paradigms of space, time, gravity, and matter while offering new ways to understand the soul, mind, and consciousness.

Sandeep's work merges logic with poetic clarity, scientific depth with philosophical courage. He is known for writing from first principles—refusing to echo authority, always starting from alignment. His books are not just texts but intellectual awakenings.

He writes not to explain the universe, but to remember it.

"I never sought fame. I sought coherence."

—Sandeep J. Chavan

Final Note to the Reader

This wasn't written to change your mind. It was written to help you listen to it better. Every page was a mirror—not for the universe, but for the subtle, echoing space where you quietly exist.

You may forget the words, but I hope you remember the sensation—that fleeting ripple when something aligned, even for a second.

Read. Understand. Feel. And then never ask— **What am I?** Because in that stillness...**You already are.**

Don't miss out!

Visit the website below and you can sign up to receive emails whenever Sandeep Chavan publishes a new book. There's no charge and no obligation.

https://books2read.com/r/B-A-EPGPC-ZOIUG

BOOKS 2 READ

Connecting independent readers to independent writers.

Did you love *Our Story*? Then you should read *Aham Brahmasmi*[1] by Sandeep Chavan!

[2]

Aham Brahmasmi – The Ripple That Remembers Itself*What if you never had to chase truth—only clear the noise that made you forget it?*

This groundbreaking work by **Sandeep Chavan** merges science, consciousness, philosophy, and ancient wisdom into a single realization: **you are not a separate self, but a ripple of the universe in alignment—or distortion.**Rooted in the paradigm-shifting framework of **Universal Energy Dynamics (UED)**, the book dismantles the illusion of individuality and

1. https://books2read.com/u/4ErjjY

2. https://books2read.com/u/4ErjjY

reframes existence as a field-based interaction. It challenges conventional beliefs, organized religion, and identity-driven spirituality—not with rebellion, but with deep coherence.

From decoding religious metaphors as early operating systems to scientifically redefining karma, soul, mind, and observer, the book guides the reader through a layered awakening. You'll explore:

Why rituals, scriptures, and gods were never the destination—only early mapsHow silence, observation, and ripple alignment replace mysticism and superstitionWhy awareness—not belief—is the real compass of a coherent lifeHow ego is turbulence mistaken for identity—and how to dissolve it without effortWhat "Aham Brahmasmi" really means in modern, field-theoretical language

Each chapter builds a ripple—from the mechanics of perception to the collapse of separation—culminating in a practical yet profound embodiment of oneness: **not mystical, not abstract, but mechanically inevitable when noise is cleared.**

This is not a book of borrowed wisdom. It is an original intellectual bridge—uniting ancient insight with futuristic science, mysticism with mechanics, and belief with consequence.

Whether you're a **seeker, scientist, philosopher, or skeptic**, *Aham Brahmasmi* invites you to pause the chase for external truth and rediscover the still ripple within.

You don't need God. You need awareness.You don't become Brahman.You remember you were never not.

If you're ready to move from worship to wisdom, from identity to interface, and from noise to ripple harmony—this is the book that will change how you see everything.

Read more at https://www.sandeepjchavan.com/.

Also by Sandeep Chavan

A Monk on World Tour
The Making of the Monk

Chavan's UED research
CMGED Research Report
Chavan's Cosmic Continuum (C^3)
Frequency
Ripples of Reality
Reality Realigned
The Great Realignment
I – The Cosmic Leverage
Chavan's Cosmic Theory of Relativity (CTR)
Chavan's Theory of Light
Feeling Like a Physicist
TIME – The New Definition
$E \neq mc^2$: The Collapse of Classical Assumptions
The Chavan Codex
Chavan's Atomic Model
Chavan's Theory of Everything

Echoes of War: The Russia-Ukraine Conflict and Its Global Ripples

Shattered Lives

The War Economy

Borders Redrawn

Arms and Allies

Collateral Damage

Justice in the Ashes

Somewhere in the Darkness of the Cosmos

The Blueprint of the Universe

The Cosmic Breath

From the Edge of the Universe

The Eternal Sage Chronicles

The Sage and the Sovereign

The Family Pillar

The People's Flame

The Warrior's Soul

The Bridging Mentor

The Guiding Light

The Exodus

The Last Cycle: 2075

The Fundamental Theory of Life & Existence

Energy in Action: The Soul Demystified

The Active Mind

Energy in Connection

Death: The Disconnection

Subconsciousness: The Chosen Disconnection

Desires: The Engine of Life

The Restoration of Human Knowledge

Restoring Bhagavad Gita

The War of Thoughts

Cracks in the Known

The Human Game

The Divine Dilemma

The Original Intent

The Answers

Ultra Modern Life Philosophy

The Duality of Laughter and Tears

Take It Easy

Compare to Accept

Standalone

It's Not AI, It's AHI - Amplified Human Intelligence
The Decision Paradox
The Four Sapiens
Malicious Script of Indian Polity
The IIT Legacy & Global Impact
The IIT Legacy & Global Impact
You are just a Version of Your Original
Win the Game You Didn't Choose
Always Keep Your Bags Packed
Reveal to Shield - Learn the Game of Facades
Engage Beyond Elections
Culture, Identity & Change: The Evolution of Indian Society
The Triangular Dynamics
Never Trust Without Doubt
If You Can't Change the Country, Leave the Country
Live Fearless with Fear
India: The Investment Magnet
Realignment
Powering a Billion Dreams
Universal Skill
Extension: From Brush to Algorithm
Human Side of Einstein
The Future Knows No Lines
Let's Bring the Generations to the Table
Unified Energy Dynamics
A Beat That Never Dies
Strategic Horizons
Right From Scratch

The Forgotten Innovators
Physics Fixing
The Savior Illusion
The Selfish Side
The Age of the Weakest Sapiens
Neural Wars
The Absorption: There Is No Death
Strange Therapist
Bubbles: Become a Drop
Beyond the Blue
Half-Scholar
The Law of Truth
Failure
Philosophy 2.0: A Structural Manifesto
Chaos: A Gentle Guide to the Unresolvable
Self-Help: New Syllabus
Rippling
Game: How to Play the Share Market
Out of Syllabus
Saffronized: The Code Behind the Crown
The Anger
Moksha: The Liberation
Chavan vs Einstein: The Universe Was Never Curved
Rocket Science: The Ripple Gambling
The Cosmos: The Still Universe
The Taste of Silence
Chavan's Post-Materialist Human Model
10-Second Rule
Chavan vs Hawking
It was Never a Wave Signal
Quantum Courtroom: The Trial of a Century

Chavan–Heisenberg Ripple Uncertainty Principle
The Great Disruption
Dark ≠ Hidden
Breath
Calculus of Life
The Earth Doesn't Move
Black Hole Don't Exist
Chavan vs Kant
Naked Truth: Everything and Else
The Moon: Co-Author of Life's Script
World War-3
Aham Brahmasmi
Chavan's Declaration
Light: Neither a Wave, Nor a Photon
Chavan vs. Plato
Good Morning
Myth Wrapped in Math
Our Story
Rethinking Energy: Ripple Tension, Not Substance

Watch for more at https://www.sandeepjchavan.com/.

www.ingramcontent.com/pod-product-compliance
Lightning Source LLC
LaVergne TN
LVHW090358210926
848480LV00002B/378

* 9 7 9 8 2 3 1 3 1 9 7 0 1 *